**LECTURES OF THE AIR CORPS TACTICAL SCHOOL
AND AMERICAN STRATEGIC BOMBING IN WORLD WAR II**

LECTURES OF THE AIR CORPS TACTICAL SCHOOL AND AMERICAN STRATEGIC BOMBING IN WORLD WAR II

EDITED AND WITH COMMENTARY
BY PHIL HAUN

UNIVERSITY PRESS OF KENTUCKY

Copyright © 2019 by The University Press of Kentucky

Scholarly publisher for the Commonwealth,
serving Bellarmine University, Berea College, Centre
College of Kentucky, Eastern Kentucky University,
The Filson Historical Society, Georgetown College,
Kentucky Historical Society, Kentucky State University,
Morehead State University, Murray State University,
Northern Kentucky University, Transylvania University,
University of Kentucky, University of Louisville,
and Western Kentucky University.
All rights reserved.

Editorial and Sales Offices: The University Press of Kentucky
663 South Limestone Street, Lexington, Kentucky 40508-4008
www.kentuckypress.com

Unless otherwise noted, photographs and maps are from the US Air Force.

The views expressed in this book are those of the author and do not necessarily reflect the official policy or position of the Department of Defense or any of its services.

Library of Congress Cataloging-in-Publication Data
Names: Haun, Phil M., editor. | United States. Army. Air Corps Tactical School.
Title: Lectures of the Air Corps Tactical School and American Strategic
  Bombing in World War II / Edited and with commentary by Phil Haun.
Description: Lexington, Kentucky : The University Press of Kentucky, 2019. |
  Includes bibliographical references and index.
Identifiers: LCCN 2018056724| ISBN 9780813176789 (hardcover : alk. paper) |
  ISBN 9780813176802 (pdf) | ISBN 9780813176796 (epub)
Subjects: LCSH: Bombing, Aerial. | World War, 1939–1945—Aerial operations,
  American.
Classification: LCC UG700 .L43 2019 | DDC 358.4/24—dc23

This book is printed on acid-free paper meeting
the requirements of the American National Standard
for Permanence in Paper for Printed Library Materials.

Manufactured in the United States of America.

Member of the Association
of University Presses

For Clayton and Sadie, who have given me hope and happiness

Contents

List of Illustrations   ix
Preface   xi
Note on the Text   xv

Introduction   1

1. Air Power and War   33
   An Inquiry into the Subject "War" *by Harold George*   33
2. The Objective of Air Warfare   46
   Air Power and Air Warfare *by Muir S. Fairchild*   46
   Principles of War *by Donald Wilson*   58
   The Aim in War *by Haywood Hansell*   72
3. The Bomber Always Gets Through   87
   Driving Home the Bombardment Attack *by Kenneth Walker*   87
   Tactical Offense and Tactical Defense *by Frederick Hopkins*   98
4. High-Altitude Daylight Precision Bombardment   117
   Practical Bombing Probabilities *by Laurence Kuter*   117
5. Vital and Vulnerable   139
   National Economic Structure *by Muir S. Fairchild*   139
   New York Industrial Area *by Muir S. Fairchild*   164
6. What to Target: The Economy or Military Forces?   180
   Primary Strategic Objectives of Air Forces *by Muir S. Fairchild*   180
7. High-Altitude Daylight Precision Bombing in World War II   195

Appendix 1. Trenchard Memo   225
Appendix 2. AWPD-1   232
Appendix 3. AWPD-42   252

Appendix 4. Combined Bomber Offensive Directive   265
Appendix 5. Pointblank Directive   267
Notes   271
Bibliography   283
Index   289

Illustrations

Formation of B-17Fs of the 92nd Bombardment Group   3
B-17F raid on Marienburg Focke-Wulf plant   4
Memphis Belle   4
Brigadier General William "Billy" Mitchell   14
US Army Air Service Martin MB-2 bomber drops a white phosphorous practice bomb on the USS *Alabama*   15
US Army Air Service Martin MB-2 bomber sinks the German battleship *Ostfriesland*   16
The 1925 court-martial of Brigadier General Billy Mitchell   18
US Army Air Service Martin MB-2 bomber   21
US Army Air Corps Curtiss B-2 Condor   22
Boeing Y1B-9A bomber and a P-26 pursuit   23
Martin B-10B bomber   23
Norden bombsight   24
Douglas B-18A Bolo   25
Boeing YB-17 Flying Fortress   25
Boeing B-17G Flying Fortress   26
Newly constructed Austin Hall, 1931   27
ACTS faculty members, 1933–1934   27
ACTS classroom   28
Harold George   34
Muir Fairchild   47
Donald Wilson   60
Haywood Hansell   73
Kenneth Walker   88
Bomber squadron in a Javelin formation   92
Bomber group in a Route Column formation   95

Laurence Kuter   118
"The Aerial Bomb vs. Public Service Electric Power"   175
"The Aerial Bomb vs. Traction Electric Power"   175
US Army Air Staff, 1941   199
AWPD-1: Air Offensive against Germany   200
B-17F formation over Schweinfurt, August 17, 1943   205
B-17F inverted over Germany after losing right wing   206
B-17 formation dropping bombs through clouds, 1945   206
Republic P-47 Thunderbolt with external drop tank   210
North American P-51 Mustang formation   211
Boeing B-29 Superfortress   218
Lauris Norstad, Curtis LeMay, and Thomas Power   221

Preface

This book began as research for a World War II lecture at the US Naval War College entitled "Victory through Air Power?" I assigned as reading Giulio Douhet's *Command of the Air*, but when it came to the American strategy of high-altitude daylight precision bombing (HADPB), I found I had no primary sources from the Air Corps Tactical School (ACTS) that had developed the theory. The following year, while conducting research at the Air Force Historical Research Agency at Maxwell Air Force Base (AFB), Alabama, I examined the files on the ACTS. I found a treasure trove of lectures that have often been cited but had never before been published. Lectures from the late 1920s and early 1930s were mostly scattered notes, but by the late 1930s, the school's leadership had begun to require the ACTS faculty to type out each lecture for review prior to being presented to the students. As a result, these lectures have been preserved exactly as they were read aloud to the US Army Air Force (USAAF) officers who would be tasked to lead their squadrons, bombardment groups, and air wings into combat.

From these lectures I assigned one, Muir Fairchild's "National Economic Structure," but without context many students and some of the faculty did not see the relevance of the reading and it was soon scrapped from subsequent syllabi. For a time I considered what format could best make the ACTS lectures accessible to military officers and scholars. The original idea was to create something akin to an edited volume in which the ACTS lectures would form the body of the book. Finally, I had the time to commence the project in the fall of 2014 while commanding the Air Force ROTC detachment at Yale University. After combing through dozens of ACTS lectures, I identified those that best presented the ACTS theory of strategic bombing.

One evening, while at a Security Studies Program (SSP) dinner at the Massachusetts Institute of Technology, I mentioned the book to the historian Frank Gavin, describing it as a history. He encouraged me to instead reconsider the project as a political scientist as I had been trained. Upon reflection, I realized that the case of Eighth Air Force's experience in the Combined Bomber Offensive against Germany presented an excellent case to test the ACTS theory of strategic bombing. The ACTS faculty, led by Harold George, Donald Wilson, and Kenneth Walker, had developed a theory of strategic bombing based on the unescorted daylight bombing of the vital nodes of an enemy nation's economy. In August 1941, the newly formed Air War Plans Division (AWPD), led by George and assisted by Walker and two other former ACTS faculty members, Haywood Hansell and Laurence Kuter, was provided the rare opportunity to test the theory in AWPD-1, the US air campaign plan for war against Germany and Japan.

In the concluding chapter of this book, I examine how these airmen turned theory into practice and assess how the bomber forces they created contributed to the Allied victory. Over the past two decades there have been several important revisionist historical works that much improve our understanding of the impact air power had on World War II. I rely heavily on these works, which include such must-reads as Tami Biddle's *Reality and Rhetoric in Air Warfare* (2002), Adam Tooze's *Wages of Destruction* (2006), Mark Clodfelter's *Beneficial Bombing* (2010), Richard Overy's *Bomber and the Bombed* (2013), and Phillips O'Brien's *How the War Was Won* (2015).

When I commenced this project, one of my questions concerned the qualifications of the ACTS faculty. What intellectual capacity, either through education or combat experience, did these airmen have to develop a theory for how air power could independently cause a modern industrial nation to lose the will and means to fight? Overall, they were well educated. All but one, Kenneth Walker, had a college education, and Harold George had a law degree. Several had technical backgrounds, including Wilson, Hansell, and Kuter. Kuter was the sole West Point graduate. While they were all experienced aviators, they had almost no combat experience. Only George had fought in World War I, but only in the last week of the war. None were trained as economists, political scientists, or historians. In addition, these officers had limited professional military education other than having attended the ACTS: Wilson had attended the Army General Command and Staff College and Fairchild the Army Industrial College and War College. Overall this was a smart group of men, but their confidence in what air power could achieve independently was supported not by any special expertise or their own combat experience but rather by their faith and belief in HADPB.

This project would not have been possible without the assistance and encouragement of a number of people. My thanks to Tim Schultz, Barry Posen, Owen Cote, Colin Jackson, Williamson Murray, Mark Clodfelter, Richard Muller, Paul Kennedy, Angus Ross, Ray O'Mara, Sally Paine, Nick Sarantakes, Marc Genest, David Stone, Joshua Rovner, Joshua Shifrinson, Archie Difante, Maranda Gilmore, Jean Carrillo, Holly Hermes, and Geoffrey Burn. A special acknowledgment is reserved for Bonnie, who is not only a phenomenal editor of my rambling prose but an even better partner for life.

Note on the Text

The body of this book consists of lectures given at the Air Corps Tactical School at Maxwell Field, Alabama, from 1936 to 1940. Records of the ACTS, including typewritten copies of lectures, are located in the Air Force Historical Research Agency at Maxwell AFB, Alabama. The ten lectures selected are from more than sixty given in the Air Force and Bombardment courses from 1932 to 1940. The lectures were chosen based on the following criteria. First, these are the most important lectures, the ones most quoted in books on US strategic bombing in World War II. Combined, these lectures lay out the logic and assumptions for a uniquely American theory of strategic bombing theory based on high-altitude daylight precision bombing.

Second, preference has been given to later lectures as these present the most mature thinking on American strategic bombing theory in the years just prior to and in the first two years before the US entry into World War II. The one exception is Kenneth Walker's 1930 *Coast Artillery Journal* article "Driving Home the Bombardment Attack." This article was chosen as a succinct summary of the 1931 bombardment textbook that Walker wrote for the ACTS Bombardment Section.

Third, these lectures were the ones presented to the largest number of ACTS students, those who attended from 1938 to 1940, and they became the officers who would soon be tasked with turning strategic bombing theory into practice over Germany and Japan.

Fourth, as a pragmatic matter, these are the lectures that have been best preserved. By the late 1930s the ACTS leadership required faculty to type out their lectures and have them approved before they were delivered to students. There is, however, a downside to using these later lectures in that not all of them were given by the original author. The best example is Donald Wilson, who is credited with developing the analytical techniques for evaluating the vital and vulnerable nodes of a nation's economic networks. The text for his 1936 lectures on "National Economic Structure" and "New York Industrial Area" were never typed out; only outlines and notes for the lectures remain. Fortunately, Muir Fairchild's later versions of the lectures by the same titles, given in 1939, are preserved and presented in the book.

Fifth, works by Harold George, Kenneth Walker, Haywood Hansell, and Laurence Kuter are included because these ACTS faculty alumni were the key air planners tasked by General Hap Arnold in early August of 1941 to coauthor AWPD-1 (Air War Plans Division), the blueprint for defeating Germany and Japan. These officers were responsible for turning theory into reality, selecting the targets and then calculating the aircraft and aircrew requirements for conducting a strategic bombing campaign.

Finally and unfortunately, there is not enough space in this volume to include all the lectures of the Air Corps Tactical School. However, the lectures presented here are primary sources for airmen, historians, and anyone interested in understanding the creation of the air power theory that underpinned the most important strategic bombing campaigns in history.

Introduction

At 0956 hours on April 17, 1943, the B-17F dubbed *Stupntakit* slowly lifted off the runway at RAF Bassingbourn, laden with five 1,000-pound high-explosive bombs destined for the Focke-Wulf plant at Bremen. *Stupntakit* was lead aircraft for the thirty-two B-17s assigned to the 91st Bomb Group, one of four groups contributing to a total of 115 bombers, Eighth Air Force's largest strike force to date. Single file, the Flying Fortresses climbed eastward as streams of bombers from the nearby bases that littered East Anglia rose to join them.[1]

In the mission brief earlier that morning, intel had informed the aircrew that the Bremen plant assembled eighty Fw-190 single-engine fighters per month, one-third of German Focke-Wulf production. The well-fortified city bristled with an estimated 500 anti-aircraft artillery (AAA or flak) batteries. Inclement weather had delayed this mission for nearly two weeks, but on this day forecasters predicted good visibility and high ceilings en route and over the target area. Throughout the winter and spring of 1943, the weather had limited high-altitude daylight operations over Germany to only four days per month, so Eighth Air Force was especially keen to take full advantage of this most rare of sunny days.[2]

Over the North Sea, the bombers circled to form up into two combat wings, each consisted of three 18-aircraft groups, then proceeded east on their seven-hour journey to Germany and back. Unfortunately, a Luftwaffe reconnaissance plane spotted the aerial armada well out over the North Sea and notified the entire coastal defense force, providing an even earlier warning than the detection time expected by ground-based early-warning radars.[3]

As the unescorted B-17s approached the Frisian Islands just off the northwest German coastline, they encountered light flak but no fighters. The large formation went "feet dry" as it crossed the coast and tracked just west of Oldenburg toward the final turn point: the old Zeppelin airfield at Ahlhorn. Then, banking hard left, the planes turned toward Wildeshausen, the initial point (IP) for the bomb run. By the IP the two wings had separated by five miles and, as the lead wing began its bomb run, 150 German fighters swarmed the formation. The enemy defenders consisted mostly of single-engine Messerschmitt Me-109s and Focke-Wulf Fw-190s, but intermixed were two-engine Messerschmitt Me-110 night fighters. Though the IP was covered in haze, the B-17s pressed on as the skies cleared over Bremen, except for the small dark clouds formed by the explosion of German 88- and 105-millimeter shells, the most intensive flak barrage Eighth Air Force had yet encountered. As the lead bombardiers squinted through their Norden bombsights during the crucial seconds prior to bomb release, the Luftwaffe fighters ignored the flak and attacked in waves to disrupt the accuracy of the weapons release.

Despite the intense defenses, many of the B-17s managed to hit their target, with post-mission analysis assessing the southern half of the Focke-Wulf complex as severely damaged by ninety-two 1,000-pound bombs. Egressing off target, the bombers turned west back toward Wildeshausen, then off to Ahlhorn before turning north to go "feet wet" over the coast near Emden. The Luftwaffe relentlessly pursued the bombers, letting up only 40 miles out over the North Sea when RAF Spitfires finally appeared.[4]

The strike force had succeeded in damaging its assigned target, but the price had been high, twice that paid on any previous mission. Eighth Air Force sacrificed 16 bombers (14% attrition), primarily to Luftwaffe fighters, and half of the returning aircraft suffered battle damage.[5] All losses were suffered by the lead wing as the Luftwaffe chose frontal attacks on the lead formations to take advantage of known limitations with the B-17F's nose gun.[6]

As for the aircrew, 159 did not return. An additionally painful blow was the loss of the crew of *Invasion 2nd* on its 24th mission. Oscar-winning director William Wyler had been filming the crew for several weeks for a war documentary on the first crew to reach 25 missions. Such heavy losses, coupled with the relentless clouds that returned to cloak Germany, forced a delay of nearly a month before American bombers could again attack the Wehrmacht. Though the Bremen raid prompted Eighth Air Force Fighter Command to request an additional twenty P-47 fighter groups, it would not be until February 1944, a full 10 months later, that long-range escorts were available.[7]

Introduction   3

Formation of B-17Fs of the 92nd Bombardment Group en route from England to Germany, circa 1943.

Compared to US bombers, the Luftwaffe suffered little. Though B-17 gunners claimed 63 enemy fighters destroyed, the actual number was five (only three to enemy fire) with another nine damaged. In addition, the setback to Focke-Wulf production was less than anticipated since, unknown to the allies, by 1943 the Germans had dispersed Fw-190 production across eight facilities. Though dispersal added to the cost of each fighter by as much as 30%, it afforded the Luftwaffe the ability to absorb the lost production from a single plant.[8]

American Strategic Bombing Theory

The conditions encountered by the B-17 raid on Bremen—weather delays, lethal enemy air defenses, and dispersed production facilities—are indicative of the broader challenges that confronted US airmen in their effort to win the

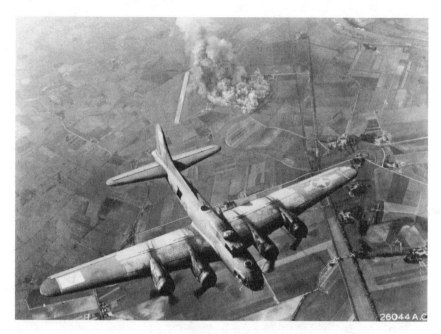

B-17F raid on Marienburg Focke-Wulf plant, October 9, 1943. (National Archives)

The *Memphis Belle*, a B-17F Flying Fortress assigned to the 91st Bombardment Group, flew her nineteenth mission against Bremen on April 17, 1943.

war with a uniquely American vision of strategic bombing. During the interwar period a small cadre of airmen serving on the faculty of the Air Corps Tactical School (ACTS) at Maxwell Field, Alabama, had articulated the concept of high-altitude daylight precision bombing (HADPB), a coherent yet controversial theory for victory through the independent employment of air forces. The ACTS instructors and their students would later be responsible for translating theory into practice in World War II. In so doing, the logic of HADPB was tested and in many ways found wanting as the advent of radar tilted the offense-defense balance toward the defenders, and the persistent clouds over Germany and Japan neutralized the potential lethality of high-altitude daylight precision air strikes.

Though not as decisive as envisioned by the ACTS, strategic bombing proved to be a critical component of the Allied victory. In early 1944, the threat from the daylight bombing of the German heartland compelled Luftwaffe fighters into the air to engage the bomber formations and their lethal long-range fighter escorts on unequal terms. US fighters had decimated the Luftwaffe by late spring, clearing the skies over the beaches of Normandy for the Allied ground invasion. Meanwhile, air interdiction missions in France prevented the German Army's strategic reserves from fortifying their defenses. As the Western Front moved eastward, US air strikes against German petroleum production and transportation networks collapsed the German economy, thus starving the German Army and Air Force of fuel and supplies. Though US strategic bombing did not win the war on its own, it was a major contributor to the combined arms effort that proved decisive in western Europe. Meanwhile, in the Pacific, a joint effort was required as an extensive naval campaign was needed to provide basing and logistics for US B-29 bombers to reach the Japanese islands. Though a number of factors contributed to the Japanese defeat, it was the B-29s that delivered the two atomic bombs to Hiroshima and Nagasaki, triggering Japan's surrender.

Even though the strategic bombing theorists were proven wrong in World War II—air forces almost never independently win wars—their ideas were critical during the interwar period in prompting the United States to develop the long-range aircraft capable of projecting combat power across continents. Without the airmen's prewar efforts, such offensive capabilities would not have existed and the Americans would have been constrained, as were the Germans and Japanese, to short- and medium-range aircraft designed to support ground forces. It was fortuitous for the United States that the Air Corps Tactical School had developed a theory for strategic bombing that justified the creation of a large air force necessary to fight a global war.

This introduction traces American strategic bombing theory from its origins in World War I to the thinking of the three great interwar air power theorists—the Italian Giulio Douhet, the Briton Hugh Trenchard, and the American Billy Mitchell—to the founding of the Air Corps Tactical School, the development of the Norden bombsight and B-17 bomber, and the genesis of HADPB theory at the Air Corps Tactical School.

Strategic Bombing in World War I

At the commencement of World War I, all the major powers had small numbers of light and unarmed airplanes assigned to observation units. As the Western Front ossified, negating the cavalry's traditional role of probing enemy defenses, the use of aircraft for obtaining battlefield intelligence proved vital. Airplanes were fitted with cameras for reconnaissance and, soon thereafter, wireless radios for artillery spotting. Airmen carried weapons, dropping grenades onto enemy positions and firing their guns at enemy aircraft. The utility of air power over the battlefield was obvious to all, as was the necessity of denying the same advantage to the enemy.[9]

Striving for command of the air spurred the development of purpose-designed pursuit aircraft. As dogfights spiraled above the trenches, the advantage shifted back and forth, with adversaries seeking technological advantages in airframe and engine design and improvements in the lethality of air-to-air weapons.

In addition to air superiority, aircraft at the front were employed for observation, reconnaissance, attack, and interdiction missions. While these missions supported the ground war, efforts were also under way to bomb the enemy's heartland. It was the German Navy that first had the capability to strike deep inside enemy territory. Zeppelin dirigibles, designed for naval reconnaissance, could travel extended distances and carry heavy bomb loads. Nighttime raids on Britain commenced in January 1915. Initially directed against coastal targets, they soon ventured inland to attack London. Though the physical damage from the raids was minor, the reaction of the British population was anything but, with cries for improved air defenses and demands for retaliatory air strikes. Zeppelins continued to threaten England until January 1916, by which time the British had developed an acceptable air defense warning network combined with pursuit aircraft capable of intercepting and strafing the slow-moving and flammable airships. Subsequent Zeppelin attacks were sporadic and ended in December 1916 following a series of losses.[10]

Respite for Britain was short-lived, however, as the Germans introduced their large two-engine Gotha Bombers and commenced daytime raids on London in June 1917. These air raids prompted a renewed outcry for further enhancement in air defenses and retaliatory attacks. Prime Minister David Lloyd George appointed General Jan Christian Smuts to assess the situation, and the Smuts Committee recommended improvement in air defenses and the founding of an independent strategic bombing force. Continued fears that Germany would escalate its strategic bombing campaign convinced the British government to create the independent Royal Air Force on April 1, 1918.[11]

British airmen had been brooding over how to bomb Germany since the first Zeppelin raids. On September 3, 1917, Major Lord Tiverton presented the first long-range operational plan to target four clusters of cities comprising Germany's key war industries. American officers had been observers within the British forces even before the United States' entry into the war in 1917. US Army Major Edgar Gorrell recycled Tiverton's plan in a November 1917 report that became the United States' first strategic bombing plan.[12]

The implementation of a long-range bombing strategy, however, proved challenging for the British and beyond the capability of the United States. In October 1917, the Royal Flying Corps No. 41 Wing was tasked with reprisal raids, but struggled throughout the winter with bad weather, unsuitable aircraft, and poorly trained pilots. Renamed the Independent Force (IF) in June, its commander, Major General Hugh Trenchard, could muster no more than a handful of bombers for raids on Germany, and these aviators had difficulty navigating their way to assigned targets and even more trouble hitting the targets they did manage to locate. By war's end, the IF had primarily bombed airfields and railroads, while only 16% of strikes were scattered across the industrial targets in western Germany within range of the British bombers.[13]

After the war the British, German, and American military staffs independently conducted bombing surveys to evaluate the impact British bombing had had on German wartime production. The British report, conducted under the watchful eye of Trenchard, concluded that while the material damage from bombing had been minor, the effect on German worker morale had been extensive. The report concluded that the threat of air raids had significantly disrupted production even at factories that had not been bombed.[14]

While the RAF handed out passing marks for its strategic bombing efforts, the assessment of the German General Staff was more critical. The Germans admitted that the IF had compelled them to redeploy air defenses for homeland defense, but the number of men, artillery pieces, and aircraft repositioned was lower than the British estimate, and this diversion of resources had not had a significant impact on German force levels available to the Western Front. While the Germans agreed with the RAF that the material damage to their factories from air raids had been limited, they concluded that the overall impact on war production and worker morale had been minor.[15]

Edgar Gorrell authored the US bombing survey, which included a section assessing British bombing damage to German factories. The report mirrored the findings of the RAF except for criticism of the IF for not concentrating attacks against key industrial targets. The RAF justified its decision to diffuse its attacks as an effort to erode worker morale across multiple cities and industries. Had the strikes been concentrated against vital economic choke points, Gorrell reasoned, the strikes would have had much greater impact. This interpretation of the British experience in World War I would lead the Americans to develop a divergent theory of strategic bombing during the interwar period.[16]

In addition to the direct lessons taken from the IF's strategic bombing campaign, the static nature of the Western Front affected the approach of military strategists as to how future wars should be fought. For many, air power promised the means to avoid the enormous costs and suffering endured by land-based attritional warfare. Also, the manner in which World War I ended with the collapse of civilian morale at home while German soldiers remained deployed on foreign soil had a deep impact on military thinking. Many soldiers and airmen on both sides of the trenches believed that the German military had been "stabbed in the back" by weak and undisciplined civilians who had lost the will to fight. A central tenet of strategic bombing theory based on this lesson was that air power could avoid another lengthy and bloody war by overflying the enemy's defenses and directly targeting the vulnerable will of the people.

Strategic Bombing Theory in the Interwar Period

In addition to the lessons taken from World War I, geostrategic concerns played a role as to which nations embraced strategic bombing. The continental powers of Germany, France, and Russia, whose survival depended on defending against invasion, focused predominantly on how to better integrate air and ground forces in combined arms approaches to warfare.[17] Italy, Great Britain, and the United States, on the other hand, countries with the luxury of having intervening bodies of water between them and their enemies, considered how air power might be employed independently. Italy was the first to produce an influential air power theorist in Giulio Douhet.

Giulio Douhet

Giulio Douhet began his career as an artillery officer when he first saw the potential for air power in 1911 during Italy's war in Libya against the Ottoman Empire. He commanded an aviation battalion in World War I, but was court-martialed for criticizing the ineptness of the Italian Army. Incarcerated for a year, he was vindicated and reinstated in 1917 after the Italian defeat at the Battle of Caporetto. By 1918, he was directing the Italian Central Aeronautic Bureau, where he worked closely with the Italian aircraft manufacturer Giovanni Caproni. In 1921, Douhet was promoted to general officer and in the same year published his treatise on strategic bombing: *Command of the Air*.[18]

Douhet proposed the employment of air power in two stages. First, a nation had to invest in an independent air force in peacetime in order to be in the position to seize command of the air in wartime. He viewed air power as an inherently offensive weapon and believed that the air force that attacked first would have the advantage of surprise, initiative, and flexibility to allow it to concentrate and attack where it chose. By contrast, a reactive air force had to defend everywhere, compelling it to disperse its forces, thus tendering the advantage to the attacker. Douhet dismissed pursuit aviation as a defensive tactic ineffective against a massed strike package of well-armed and armored battleplanes flying a tight, mutually supportive formation. By going on the offense, an independent bomber force could gain and maintain command of the air by immobilizing the enemy's nascent air force in its factories, effectively "destroying eggs and the nests."[19]

After gaining command of the air, the next phase to winning the war was to crush the enemy's resistance. To Douhet, a successful air strategy depended on target selection. In the next war, which he anticipated would be total, entire nations would be mobilized. He argued that, under such conditions, there would be no moral or legal distinction between the military and civilians. Air power, unlike surface forces, could avoid the adversary's defenses and directly attack the enemy's vulnerable civilian population. Attacking the population would weaken the structure of society and shatter its morale. Having lost the will to fight, the people would then demand their government put an end to war. Douhet also argued that by directly bombing civilians, the war would end more quickly and thus result in less bloodshed than long, drawn-out attritional warfare.[20]

Three criticisms of Douhet's theory are of note. First, his faith in the offensive nature of air power, central to his strategic bombing theory, later proved misplaced. In World War II, the advent of radar and technological advances in communications and the improvements in the performance of pursuit aircraft would reduce or negate the advantage he predicted for an attacking force. Effective air warning systems developed by Britain and Germany meant that defenses could concentrate fighters against incoming bombers. Though the bomber force still got through to the target, without gaining air superiority first, the attrition suffered would prove too high for sustained operations. Second, Douhet overestimated the fragility of civilian morale and underestimated the degree of control such powerful governments as those of Great Britain, Germany, and Japan would exert over their population. Finally, Douhet did not foresee how air power might be used as a deterrent to war. If adversaries, each having independent air forces, lacked the means to defend against devastating air strikes, as with the nuclear stand-off between the United States and the Soviet Union during the Cold War, then an incentive for restraint would arise on both sides, rather than the preemptive use of force his theory anticipated.[21]

In defense of Douhet, there was no way he could have anticipated how radar would shift the advantage in air power from the offense to the defense. Moreover, despite the advantage of radar, pursuit aircraft could only attrit and curb, but not prevent, a determined bomber force from reaching its target. In the strategic calculations of the Cold War, the United States and Soviet Union both counted on the offensive superiority of their bombers and intercontinental ballistic missiles to make the threat of nuclear strike credible. Second, though Douhet may have misunderstood the control the German, Japanese, and British governments could exercise over civilians, he fully grasped the tenuous relationship between Mussolini and the Italian population as evidenced by Mussolini's ouster, albeit temporarily, after the American bombing of Rome in July 1943.[22]

Overall, the enduring value of Douhet's theory lies in his articulation of the importance of gaining command of the air and in his attention to target selection. While he may not have predicted the degree to which air defenses would neutralize offensive forces and he underestimated the difficulty of collapsing the enemy's will by bombing its civilians, his advocacy for an independent air force "in being" prior to the onset of hostilities, capable of immediately gaining command of the air and striking vital enemy targets, endures.

The degree to which Douhet directly influenced the thinking of American airmen at the time is debatable. US Army Air Service Brigadier General Billy Mitchell did meet with Douhet's colleague Caproni in 1917 and again in 1922. By 1923, translated summaries of *Command of the Air* were available within the Air Service. It was the British, however, and especially the father of the RAF, Hugh Trenchard, who had the greatest impact on US airmen during and in the immediate aftermath of World War I.[23]

Hugh Trenchard

Major General Hugh Trenchard commanded the Independent Forces in France when the first American airmen began arriving in Europe. Like Douhet, Trenchard believed air power to be offensive in nature. As a result, Trenchard pushed his men and equipment hard in battle, resulting in heavy losses of British airmen and aircraft. Nonetheless, Trenchard's offensive spirit endeared him to the commander of the British Expeditionary Forces, Field Marshal Douglas Haig.

Following the war, Trenchard commanded the RAF as the Chief of the Air Staff from 1918 to 1929 where his shrewd bureaucratic maneuvering managed to keep the RAF independent. From this position, he exerted enormous influence on the development of British strategic bombing theory. Trenchard concurred with Douhet that gaining command of the air was the first task of air power, a capability that required an independent air force "in being." In a fascist Italy, Douhet was unconstrained by moral and legal considerations regarding the bombing of civilians, whereas Trenchard operated within the antiwar climate of a liberal and democratic Britain. Trenchard therefore remained elusive as to the targeting specifics of a strategic bombing campaign, instead offering abstract references to attacking the enemy's morale.

The best written expression of Trenchard's thoughts on strategic bombing is found in a 1928 memo to the British Chiefs of Staff wherein he discusses the necessity for striking the vital war industries of the enemy even if this meant civilian collateral damage:

> Among military objectives must be included the factories in which war material (including aircraft) is made, the depots in which it is stored, the railway termini and docks at which it is loaded or troops entrain or embark, and in general the means of communication and transportation of military *personnel* and *material*. Such objectives may be situated in centres of population in which their destruction from the air will result in casualties also to the neighbouring civilian population, in the same way as the long-range bombardment of a defended coastal town by a naval force results also in the incidental destruction of civilian life and property. The fact that air attack may have that result is no reason for regarding the bombing as illegitimate provided all reasonable care is taken to confine the scope of the bombing to the military objective. Otherwise a belligerent would be able to secure complete immunity for his war manufactures and depots merely by locating them in a large city, which would, in effect, become *neutral* territory—a position which the opposing belligerent would never accept. What is illegitimate, as being contrary to the dictates of humanity, is the indiscriminate bombing of a city for the sole purpose of terrorizing the civilian population. It is an entirely different matter to terrorise munition workers (men and women) into absenting themselves from work or stevedores into abandoning the loading of a ship with munitions through fear of air attack upon the factory or dock concerned. Moral effect is created by the bombing in such circumstances but it is the inevitable result of a lawful operation of war—the bombing of a military objective.[24]

For Trenchard, strategic air strikes did not have to physically destroy the enemy's wartime economy to be effective. He claimed that "the morale effect of bombing stands undoubtedly to the material effect in a proportion of 20 to 1."[25] Bombing factories, or even the threat of bombing, would drive out workers and halt production. Even near misses would have an impact as, presumably, local housing included those of the workers. Trenchard concluded that sustained strikes on the enemy's war industry would break the morale of the workers and cause the collapse of the economy and the will of the people to continue the war, as he believed had happened in Germany, helping to bring about an end to World War I.

The RAF's emphasis on the offensive nature of air power and the morale effect of strategic bombing would later contribute to tragic consequences in the lead-up to World War II. Air power advocates in the interwar period had succeeded in convincing the British and French populations and their politicians of the destructive force of strategic bombing. In 1932, the future prime minister Stanley Baldwin infamously declared in the House of Commons that "the bomber would always get through."[26] Later, the fear of the Luftwaffe and the threat of air strikes on London and Paris factored into the British and French decisions to appease Hitler at Munich in 1938.[27]

Trenchard's focus on attacking morale and the de-emphasis on material damage relieved the RAF of the most pressing operational and tactical challenges for long-range bombing: navigation, target identification, and bombing accuracy. When war came, the RAF again had difficulty in locating targets and in hitting those it did identify, as documented in the infamous August 1941 Butt Report, which showed that less than a third of aircrews were able to drop bombs within five miles of their intended targets.[28]

In Trenchard's defense, focus on the morale effect of strategic bombing had several advantages for his fledgling service. First, unlike material damage, it was much more difficult to quantify the morale effect of bombing. This made it difficult for critics to refute RAF claims as to the effectiveness of strategic bombing. Also, at the operational level, if placing the bomb close to a factory was considered to be just as good as destroying a factory, then some of the thornier technical challenges of navigation, target identification, and precision bombing could be ignored. Second, Trenchard's top priority was keeping his newly formed air force "in being." Strategic bombing offered an argument for an air force capable of achieving independent political outcomes. Beyond the impact of his leadership upon the RAF, Trenchard's ideas also influenced American airmen, the most prominent of whom was Brigadier General William Mitchell.

Billy Mitchell

William "Billy" Mitchell, the flamboyant son of a US senator, joined the US Army Signal Corps in 1898. In May 1917, Major Mitchell was dispatched as an aeronautical observer to Europe, where he witnessed the technological advances to French and British aircraft design, innovation accelerated by

Brigadier General William "Billy" Mitchell, circa 1921.

the requirements of the war. It was here that Mitchell met Trenchard, who influenced his thinking on the need for an independent air force. When the American Expeditionary Force arrived later that year, Mitchell commanded the US air component, where he proved an able combat commander and gained invaluable insights into air operations. Experience convinced him that air power had a role to play equal with land and sea power and that command of the air was critical to the success of air, sea, and land warfare. Colonel Mitchell, however, returned to an isolationist America rapidly demobilizing. His call for an independent air force fell on deaf ears.[29]

Yet despite demobilization, the importance of air power was not entirely lost on the War Department. In 1920, the Army Reorganization Act formed the US Army Air Service with Brigadier General Mitchell as its assistant chief. The establishment of the Air Service did not mean, however, that army leadership believed air power should be employed independent of ground forces. Given the army's view of air power as a supporting arm, along with America's retrenchment into isolationism, Mitchell could not openly advocate for an offensive strategic bombing force designed to strike distant

US Army Air Service Martin MB-2 bomber drops a white phosphorous practice bomb on the USS *Alabama* during exercises on the Chesapeake Bay, September 23, 1921. (US Naval History and Heritage Command)

enemies. Instead, he took a different tack by warning of the threat of such an attack on the United States. He reasoned that an independent American air force could more effectively and efficiently provide for the nation's defense than could the navy's surface fleet, which Mitchell assessed as antiquated.[30]

In July 1921, Mitchell seized the opportunity to back his claims by orchestrating the well-publicized aerial bombardment demonstration that sank the German battleship *Ostfriesland* off the Virginia Capes. In January 1922, Mitchell departed on a lengthy trip to Europe: in Italy he met with Caproni, in Germany with Ernst Udet and Hermann Goering, and in Great Britain with RAF leaders. He returned to the United States even more convinced of the need for an independent air force and directed his energy toward further developing Air Service tactics and doctrine.[31]

Mitchell circulated within the Air Service a manual on air operations entitled "Notes on the Multi-motored Bombardment Group." This document best captures his thinking on strategic bombing at the time.[32] Mitchell, like

A US Army Air Service Martin MB-2 bomber sinks the German battleship *Ostfriesland* during an aerial demonstration off Cape Henry, Virginia, on July 21, 1921.

Douhet and Trenchard, believed in having an independent air force "in being" at the outset of war, capable of immediate offensive action to gain command of the air. He valued pursuit aviation, however, for both defensive and escort roles. While he publicly argued for the development of air power for coastal defense, within the Air Service he developed and shared his ideas on offensive air operations. Unlike Douhet, who advocated the direct attack of civilians, or Trenchard, who targeted worker morale, Mitchell's theory was aimed at the material destruction of the enemy's forces, from military factories to fielded forces. The chapter entitled "Attack of Land Establishments" listed the potential objectives for air attack: "The Bombardment Group Commander will receive his mission from the next higher Air Force Commander. His objective may be enemy airdromes, concentration centers, training camps, personnel pools, transportation centers, whether rail, road, sea, river or canal, ammunition and supply dumps, headquarters of staff commands, forts and heavily fortified positions, trains, convoys, columns of troops, bridges, dams, locks, power plants, tunnels, telephone and telegraph centers, manufacturing areas, water supply and growing grain."[33] Mitchell thus provided a list of potential military and civilian targets for strategic bombing, though he did not provide a theoretical framework for determining the priority in which these targets should be struck or how these attacks would lead to victory.

With regard to attacking civilians, Mitchell indicated that "it may be necessary to intimidate the civilian population in a certain area to force them to discontinue something which is having a direct bearing on the outcome of the conflict. In rare instances Bombardment aviation will be required to act as an arm of reprisal."[34]

To justify civilian casualties in manufacturing areas, Mitchell, unlike Trenchard, focused on the material impact such attacks would have in shortening war.

> The ethics of attack on manufacturing areas in the rear is one which is engaging the attention of an International Tribunal convened at the Hague. The result of the conference will probably be the foundation of a set of rules limiting attacks of this nature. Manufacturing centers produce munitions of war. Munitions of war are imperative to the continuance of the struggle. The destruction of manufacturing and material brings the conflict to a quicker termination. These facts cannot be controverted. Thus, manufacturing centers become military objectives, even though they house non-combatant personnel. If given sufficient warning that the center will be destroyed, non-combatant personnel will be unable to justify their remaining in the vicinity when the actual bombardment takes place.[35]

As to his advocating the targeting of a nation's water and grain supplies, Mitchell ignored the effect this would have on the civilian population, choosing to focus instead on the impact such loss would have on enemy's capacity to wage war and on how such attacks would shorten the conflict.[36]

In 1923, Mitchell toured the Pacific and Asia. He returned in the summer of 1924 with a prescient trip report concluding that war between the United States and Japan was inevitable, that the Philippines, Guam, and Hawaii were vulnerable to attack, and that the United States should restructure its defense posture in the Pacific with long-range, land-based bombers.[37]

Back in Washington, Mitchell continued his attack on the Navy and War Departments in a series of articles in the *Saturday Evening Post* and in congressional hearings. In response, Secretary of War John Weeks did not reappoint Mitchell as assistant Chief of the Air Service. Instead, Mitchell reverted to his permanent rank of colonel and was reassigned to Fort Sam Houston, Texas, in March 1925.[38]

In August 1925, Mitchell published *Winged Defense*, a hastily prepared book comprised of previously published articles that directly attacked Secretary Weeks. In September, following the navy's back-to-back loss of a PN-9 sea plane and the dirigible *Shenandoah,* Mitchell issued a statement calling the Navy and War Departments incompetent and criminally negligent in their administration of national defense. President Calvin Coolidge responded by appointing the Morrow Board to assess the development of air power. Coolidge further charged Mitchell with insubordination, resulting in a publicized court-martial and the resignation of his commission in February 1926.[39]

Mitchell would continue to advocate publicly for air power, but with his direct connection to the Air Service now severed, it would be left up to his lieutenants at the Air Corps Tactical School to mature American strategic

The 1925 court-martial of Brigadier General Billy Mitchell.

bombing doctrine. Unlike Douhet and Trenchard, Mitchell argued for a balance force of bombers and pursuit aircraft until the early 1930s when he, too, began to fall under the spell of the promise of unescorted long-range bombers.

AIR CORPS TACTICAL SCHOOL

Back when Mitchell was waging his air power crusade against the army and navy, the Air Service organized the Field Officers' School at Langley Field, Virginia, to instruct officers on air tactics. The school opened in November 1920, but its first academic year was cut short when Mitchell appropriated the faculty and students in the spring of 1921 to participate in the sinking of the *Ostfriesland*. In 1922, the school was renamed the Air Service Tactical School. Its commandant, Major Thomas Milling, and his assistant, Major William Sherman, developed a nine-month curriculum that included air tactics as well as more generic staff duties. In 1923, the course added a flying program with morning classroom instruction followed by the practical application of the lessons in the air in the afternoon.[40]

In July 1926, the recommendations of the Morrow Board led Congress to redesignate the Air Service as the Air Corps, giving the school its new name: the Air Corps Tactical School. Congress further authorized the expansion of the Air Corps over five years, though the Great Depression severely constrained the funds made available to do so.[41]

Though airmen envisioned the independent employment of air power, army doctrine remained conventional in its dictates. Army Training Regulation 440-15, "Fundamental Principles of Employment of the Air Service" (January 1926), had declared the purpose of air power as strictly the support of land forces. Within the Air Corps, there grew a disconnect between official army doctrine and what ACTS was teaching, particularly with regard to bombardment. The strategic bombing theory that emerged during these early years of ACTS is best expressed by Major William Sherman, who wrote the first bombardment text for the school, publishing it as a book titled *Air Warfare* in 1926.[42]

Sherman, like Douhet, Trenchard, and Mitchell, believed that, unlike surface forces, air power was inherently offensive in nature, with victory the result of the application of superior force at the decisive time and location. On the offense, the attacker had the advantage of choosing the time and place for battle, while the defender had the burden of operating an air defense system over a wide territory that made it difficult to react in time and with adequate force.[43]

Sherman diverged from Douhet, however, by anticipating that nations might be deterred, at least initially, from bombing civilians for fear of retaliation. Instead, Sherman focused on the nature of modern militaries and the vital supply system required to feed, arm, and supply them. Bombers could overfly enemy forces and attack this supply chain, the most logical target being enemy industrial centers. Unlike Trenchard, who would target the morale of the workers through dispersed area bombing, and Mitchell, who made no distinction as to which targets were most important, Sherman prioritized the targeting of key facilities that could paralyze an enemy's military production. He described industry as a "complex system of interlocking factories, each of which makes only its allotted part of the whole.... Accordingly in the majority of industries, it is necessary to destroy certain elements of the industry only, in order to cripple the whole.... On the declaration of war, these key plants should be made the objective of a systematic bombardment."[44]

Sherman's ideas were not entirely original, as the targeting of key industries could be traced back to Gorrell's strategic bombing plan from World War I. Still, this notion of air power independently targeting the vital and vulnerable nodes of an enemy nation's military industrial complex would form the central tenet of the ACTS strategic bombing theory.

ACTS struggled during the 1920s from a lack of resources and qualified faculty. As a result, only 217 officers graduated from the school while it operated at Langley Field from 1921 to 1931. Only with the expansion of the Air Corps in 1927 did class size begin to increase, going from 15 graduates to 39 by 1931. Likewise, Langley Field could not physically accommodate both ACTS and the new flying squadrons that were being formed. A new location for the school was identified near Montgomery, Alabama, and ACTS moved to its new home at Maxwell Field in July 1931.[45]

In 1928, the ACTS curriculum expanded to include the Air Force Course as a capstone event for the year. By 1935, the development of this course had become the chief concern of the Air Tactics and Strategy Department, the most important department at ACTS. The faculty produced foundational lectures that articulated the American strategic bombing theory based on HADPB. To make the dream of HADPB a reality, however, would require a long-range bomber capable of accurately dropping its bomb loads from high altitude, a quest met only by the development of the Norden bombsight and the Boeing B-17 Flying Fortress.[46]

Development of the Norden Bombsight and the Boeing B-17

During the 1920s and early 1930s, while the Air Corps Tactical School lectured on how long-range bombers would win wars by conducting independent strategic bombing campaigns, the reality was that the Air Corps flew the Martin MB-2, a primitive twin-engine biplane with a top speed of 100 knots and service ceilings of no more than 13,000 feet. In 1928, the selection of the Curtiss B-2 modestly improved performance with a speed of 115 knots, service ceiling of 17,000 feet, a bomb load of 4,000 pounds, and a combat range of 400 miles. A significant improvement would not take place until 1930 with the development of the all-metal, twin-engine monoplanes in the form of the Boeing Y1B-9 and the Martin B-10 bombers. The Martin B-10 had a top speed of 185 knots, a service ceiling of 21,000 feet, and a nautical mile combat range of 500 miles.[47]

US Army Air Service Martin MB-2 bomber, circa 1922. (US Air Force National Museum)

US Army Air Corps Curtiss B-2 Condor, circa 1930.

As these aircraft gradually expanded the flight envelope for bombers in terms of speed, range, and altitude, the Air Corps was simultaneously working to acquire the bombsights required for accurate, high-altitude weapons delivery. Interestingly, it would be the navy that succeeded in procuring such a bombsight. In 1920, the Bureau of Ordnance discussed the possibility of a gyro-stabilized bombsight with Mr. Carl L. Norden. Norden, a former engineer with the Sperry Gyroscope Company, had previously worked on various projects for the navy. The navy was interested in a bombsight capable of hitting a moving ship and the Bureau of Ordnance requested that Norden develop a general utility bombsight. For nine years Norden designed and refined his bombsights; by 1929, he had the initial design for a new bombsight, a synchronized device that promised to be accurate for high-altitude bombing. By 1931, Norden had delivered to the navy its first such device, and the following year, the bombsight proved itself superior in testing as the only sight capable of accurate bombing up to 18,000 feet.[48]

In 1933, the US Air Corps received its first Norden bombsight for testing and soon awarded contracts for production. Norden used the testing data obtained from the navy and Air Corps and, by 1935, adjustments had been incorporated into the basic design that would later be used in World War II. The following year, the Air Corps procured 100 bombsights to be mount-

Boeing Y1B-9A bomber and a P-26 pursuit, circa 1933.

Martin B-10B bomber, mid-1930s.

Norden bombsight, 1940s.

ed in the new two-engine Martin B-10s. It would eventually be in the B-17, however, that the Norden bombsight fulfilled American airmen's dreams of a high-altitude daylight precision bomber.[49]

The Air Corps' dream for a long-range strategic bomber resulted in the development of the game-changing four-engine Boeing B-17, which arrived in 1935. The US Army, however, forced the Air Corps to buy the Douglas B-18 Bolo instead, a twin-engine bomber deemed more suitable for supporting ground forces. Despite the setback, the Air Corps succeeded in procuring thirteen YB-17s designated as "test" aircraft. These aircraft formed a single squadron, which would remain the only B-17 squadron until production ramped up in 1939. By war's end, 12,700 B-17's would be built.[50]

The early B-17s had a crew of six although by World War II, the number of crew had expanded to 10. It had a top speed of 256 mph at 14,000 feet and a service ceiling of 30,000 feet. Its combat load was 4,000 pounds for long missions but could be adapted for shorter missions and larger loads by carrying less fuel. For defense, the B-17 was initially armed with five 30-caliber machine guns, with 50-caliber guns added later.[51]

By 1938, the Air Corps calculated that the B-17 mounted with the Norden bombsight had a mean error of 241 feet from 15,000 feet. This meant that half of the bombs could be expected to hit within a 241-foot radius of the target (combat experience would show these calculations to be overly optimistic).

Douglas B-18A Bolo, late 1930s.

Boeing YB-17 Flying Fortress, 1935.

Boeing B-17G Flying Fortress, circa 1943–1945.

To compensate for this dispersion, the Air Corps tactic was to drop a larger number of small bombs, the 500-pound bomb being the preferred size.[52]

The arrival of the B-17 and Norden bombsight had finally made material the Air Corps' dream of a long-range bomber that could, in turn, make the ACTS theory of high-altitude daylight precision bombing a reality.

The ACTS Faculty and Students

By 1935, the ACTS faculty at Maxwell Field had grown both in number and quality. Twenty-two officers were now assigned, including five from the other arms and services. With graduates of ACTS, the Army Command and General Staff College, Army War College, Army Industrial College, and the Naval War College, the faculty was well versed in air power theory and general military strategy.[53]

The key faculty members responsible for the development of strategic bombing theory included Harold George, the ringleader of a small group

Newly constructed Austin Hall in 1931, the home of the Air Corps Tactical School. (Air University Directorate of History)

ACTS faculty members, 1933–1934, including Donald Wilson (*first row, third from right*), Harold George (*second row, far left*), and Claire Chennault (*third row, second from left*). (Air University Directorate of History)

ACTS classroom, 1930s. (Air University Directorate of History)

of officers collectively known as the "Bomber Mafia." George directed the Air Tactics and Strategy Department from 1934 to 1936, the most influential department at ACTS. His rival was Claire Chennault, chief of the Pursuit Section, who passionately upheld the relevance of pursuit aircraft until his departure in 1935. Joining the faculty the same year as Chennault's departure, Haywood Hansell and Laurence Kuter proved able lieutenants for Harold George. Donald Wilson replaced George in 1936, serving as director of Air Tactics and Strategy until the school disbanded in 1940 as the Air Corps was rapidly expanding in preparation for war. Muir Fairchild, a bright officer (future four-star) and persuasive lecturer, became chief of the Air Force Section within the Air Tactics and Strategy Department from 1937 until the school closed in 1940.[54]

With the move to Maxwell Field back in 1931, the number of students attending ACTS had more than quadrupled, for a total of 870 officers graduating from 1931 to 1940. Nearly half attended a truncated twelve-week course from 1939 to 1940 when HADPB dominated the curriculum. ACTS graduates would go on to contribute significantly in World War II. Two-thirds of the general officers in the US Army Air Forces (USAAF) were alumni, including eleven of its thirteen three-star generals and three of its four four-star generals.[55]

HIGH-ALTITUDE DAYLIGHT PRECISION BOMBING THEORY

By the mid-1930s, the Department of Air Tactics and Strategy at ACTS had begun to offer courses on the air force, pursuit, bombardment, attack, observation, and naval operations. As a result of the Great Depression, however, the Air Corps' limited acquisition budget had compelled it to favor investment in the bomber at the expense of the fighter. This decision, along with the investment in civilian aviation for long-range transport aircraft, had, in turn, produced a technology gap further favoring the bomber's offensive capabilities over fighter defenses. It proved difficult for fighter pilots like Chennault to convince Air Corps leaders of the viability of high-performance interceptor aircraft that did not yet exist when airmen were already flying the remarkable B-17 at previously unfathomable speeds and altitudes, and dropping bombs with their new Norden bombsights with amazing accuracy.

Upon the departure of Chennault, the Bomber Mafia monopolized the school's thinking on the operational employment of air forces, placing ever-increasing emphasis on unescorted strategic bombers. It was in the Air Force Course, the capstone event at ACTS, that the faculty presented its theory of HADPB.[56] Harold George commenced the Air Force Course with "An Inquiry into the Subject 'War'" (chapter 1). The introductory lecture articulated the central thesis of the airplane as a new means of warfare whereby the heart of a nation could now be directly attacked without first having to defeat enemy military forces, a role that would make air power decisive in the future.

Subsequent lectures (chapter 2) expanded on George's thesis. Muir Fairchild, in "Air Power and Air Warfare," emphasized the need for an air force in being at the outset of war. In "Principles of War," Donald Wilson argued that instead of seeking out immutable principles, the study of war should instead focus on linking the desired ends to the available means. The true objective of war was not the defeat of the enemy's armed forces in battle but the overall defeat of the enemy nation such that it sued for peace on favorable terms. Air power was best suited for achieving this objective through massed offensive action. Haywood Hansell, in "The Aim in War," reasoned that war was the means for a nation to achieve its objectives of prosperity, security, and racial unity (in the case of Nazi Germany) by overcoming the will of the enemy nation's population. The airplane, unlike surface forces, could directly strike the nation's will without having to first defeat the enemy's military forces.

To support the assertion that air power was inherently offensive, Kenneth Walker, in "Driving Home the Bombardment Attack" (chapter 3), argued that in the air, offense dominates defense, and that a well-armed and well-flown massed bomber formation could defend against any air-to-air attack. In "Tactical Offense and Tactical Defense," Frederick Hopkins assessed future bomber attrition rates by taking an inductive, historical approach. In World War I, only when German defenders concentrated their fighters to British bombers at a ratio of 1.5 to 1 did British loss rates become too great for sustained operations. It was unlikely such ratios could be achieved in the future given the defender's dilemma of having to defend everywhere yet also mass forces against an offensive force that could choose the time and location of attack.

The operational requirements for conducting high-altitude daylight precision bombing are examined in "Practical Bombing Probabilities" (chapter 4). Laurence Kuter reviews the detailed planning required to determine the number of bombers and bombs to assign to a target in order to be reasonably confident of success. For high-altitude operations, the accuracy of the bombsight was the key factor in such calculations.

Muir Fairchild took up the critical question as to what to bomb in two lectures, "National Economic Structure" and "New York Industrial Area" (chapter 5). Targeting a relatively small number of vital and vulnerable nodes of an industrialized nation would produce economic and social paralysis, thereby quickly draining the enemy population of the will and means to resist. By contrast, Fairchild argued that direct bombing of civilians was not only unethical and diplomatically costly but also less effective than attacking a nation's infrastructure. The attack of economic targets would exert pressure both on the population's ability to sustain itself and on a country's capacity to produce the weapons, munitions, and supplies required to wage war.

In the final lecture of the Air Force Course, "Primary Strategic Objectives of Air Forces" (chapter 6), Fairchild concluded that the fundamental decision for air warfare was whether to attack an enemy's national economic structure or its military forces. Attacking the economic structure was a purely offensive action, while attacking enemy forces had both offensive and defensive objectives. In modern war, technologically sophisticated weapons played an increasingly important role, and destroying the means of producing such weapons prevented the arming of the enemy's military. The targeting of the enemy nation's economic structure thus denied the opponent the hope of victory and increased the costs of continued resistance.

Combined, these ACTS lectures presented a uniquely American theory of strategic bombing that would soon be tested in World War II. The lectures on the requirements for conducting high-altitude daylight precision bombing offer not so much original thought by the ACTS faculty as a reflection of how the Air Corps thought corporately about how to conduct air operations. The originality is found in the connection between means and ends, by articulating the causal logic of how HADPB operations would lead to victory, why the direct attack on a limited number of economic targets would curtail both the enemy's will and its capability to resist.

The rapid expansion of the Air Corps in 1940 in anticipation of war brought an end to the school and, had this been the end of the story, ACTS would be an inconsequential footnote in air power history. In August 1941, however, the ACTS Bomber Mafia, including George, Hansell, Kuter, and Walker, created AWPD-1 (Air War Plans Division), the US air campaign plan to defeat Germany and Japan. Central to the plan was an independent strategic bombing campaign based on high-altitude daylight precision bombing.

In the chapters that follow, lectures from the Air Corps Tactical School are presented as they were given to the airmen who would soon lead the USAAF in World War II. Fortunately for us, ACTS policy in the late 1930s required instructors to type out each lecture and have the transcript approved before its presentation. As a result, these lectures, given at a time when the Air Corps had matured its thinking on strategic bombing, have been preserved word perfect, exactly as recited to the future combat leaders of the USAAF. They have never been published heretofore and, unlike the secondary sources written after World War II, collectively introduce ex ante evidence of Air Corps thinking on strategic bombing during the interwar period. These lectures provide an invaluable resource for air power theorists and practitioners, as well as for security study scholars, to better understand the views of American airmen on the role of air power in World War II.

The following chapters have been gleaned from dozens of lectures given over the years at the Air Corps Tactical School. In chapters 1 to 6, nine lectures and an article present a uniquely American vision of strategic bombing by high-altitude daylight precision bombing. The airplane had changed the nature of war. An independent air force in being during peace would be decisive in war by massing formations of self-defending bombers to overfly defenses and precisely strike the enemy's vital and vulnerable economic nodes. The strategic paralysis that would ensue would prevent the enemy from the production of the materials required to wage modern warfare and deprive civilians of the goods necessary to survive. Unable to defend or feed itself, the enemy population would soon lose the will to resist. Chapter 7 assesses how well this theory of HADPB held up to the reality of combat by examining US air war planning in World War II and evaluating American strategic bombing effectiveness.

1

Air Power and War

AN INQUIRY INTO THE SUBJECT "WAR"

Major Harold George, the leading intellect of the ACTS bomber cadre, commenced the Air Force Course in 1936 with "An Inquiry into the Subject 'War,'" in which he introduces the controversial topic of whether air forces can win wars independently.[1] He considers whether the advent of the airplane has changed the very nature of war or simply added a new weapon to the arsenal. Nations once fought with only armies and navies, where victory over the enemy's forces was a necessary intermediate objective, an obstacle, the removal of which was required to overcome the enemy's will to resist.[2] George points out that modern civilization has made it advantageous to change how wars are waged. He argues that an industrial state is internally linked by a series of economic nodes vulnerable to disruption and concludes that air power can now attack the heart of a nation without it being necessary to first fight a war of attrition. In future wars, the air phase will be decisive.

An Inquiry into the Subject "War"
Major Harold George, Director, Department of Air Tactics and Strategy
1936

We now begin a phase of study which should give at least an introduction to air power and the strategy of the employment of air forces.[3] So far in our search for knowledge we have not been required to start with a fundamental principle or concept and then, by a process of reasoning and logic, deduce conclusions which, although not yet supported by actual proof, appear to

Harold George, 1933.

be reasonable, practicable, and logical. To date, your air instruction has been centered around subjects with which you already have a considerable acquaintance. You have been asked to believe merely that, with certain types of equipment, and with certain degrees of training, it is possible for an airplane, or a group of airplanes, to fly to a certain point and accomplish certain military missions. You have been asked to assemble the facts which you already know are true and make therefrom a completed picture. You all know, from personal experience, or from happenings to which you can give judicial notice, that bombs can sink powerful naval vessels and destroy such structures as reinforced concrete bridges. You know that blind flying and blind landing are possible and you know that an airplane can be navigated with accuracy over long distances. You may not have believed all the tactics or techniques taught at this school thus far but, believing or not, you have been able to subject the ideas presented to you to the analysis of personal knowledge. Insofar as your instruction in ground subjects has been concerned, you know that the tactics taught have been proven on the field of battle; that regardless of whether or not they are possible and feasible today, they were practical in the past.

From today on, much that we shall study will require us to start with nothing more than an acknowledged truth and then attempt, by the utilization of common sense and logic, to evolve a formula which we believe will stand up under the crucial test of actual conditions. We shall attempt to develop, logically, the role of air power in future wars, in the next war. We are not concerned in fighting the past war;—that was done 18 years ago. We are concerned, however, in determining how air power should be employed in the next war and what constitutes the principles governing its employment, not by journeying into the hinterlands of wild imaginings but by traveling the highway of common sense and logic.

In pursuing this purpose, we realize that air power has not proven itself under the actual test of war. We must also realize that neither land power nor sea power has proven itself in the face of modern air power.

The question for you to consider from today onward, to have constantly before you as you continue your military careers, is substantially this: "Has the advent of air power brought into existence a method of prosecution of war which has revolutionized that art and given to air forces a strategical objective of their own, independent of either land or naval forces, the attainment of which might, in itself, accomplish the purpose of war; or has air power merely added another weapon to the waging of war which makes it in fact only an auxiliary of the traditional military forces?" The correct answer to that question will color and influence the entire art of war. It is not merely a minor problem, it is of major importance; it is basic; it is fundamental because it is concerned with the strongest of man's instincts: that of self-preservation.

Now such a question can be answered intelligently only after we have inquired into the very depths of the philosophy of war. We must determine:

What is war?
Why does war occur?
What is the object of war?
How has it been waged in the past and why has it been waged in that manner?
Is it to the advantage of civilization to change the method of waging war if such a change is possible?

Has modern civilization reduced or increased the vulnerability of nations?

Have science and invention provided a new method of waging war?

After we have analyzed and discussed those questions we shall have prepared the foundation which will enable us to make an intelligent answer to our question.

If there is one thing which we desire studiously to avoid in our inquiry into the strategy of the employment of air forces, it is dogmatism. What we intend to discuss in this room are not our unsupported opinions. The conclusions which we draw here have been obtained not by closing our eyes and pulling them out of the thin air, but by study, by research, by finding out what military students throughout the world, who have devoted their lives to the subject of air power, believe to be reasonable. If what we have to offer does not appear to be sound, reasonable, and logical to you gentlemen, then it must be faulty—it must be unsound; if it cannot stand up under a searching analysis by seventy officers who earnestly desire to ascertain the truth, then it must be wrong. And, gentlemen, it is the truth which we are seeking; it is the truth which we desire.

The basic question with which we are confronted is national defense. If in our search for the truth we find that air power is merely an adjunct, an auxiliary, of land and sea power, and that its doctrines, the strategy of its employment, must have as their ultimate purpose the furtherance of the mission of those forces, then that is what we earnestly desire to ascertain. If, on the other hand, common sense and logic indicate that air power is not merely an auxiliary of the traditional instrumentalities of war, but is a separate instrumentality, that it has a role in future warfare separate and distinct from land or sea power, a role which it alone can accomplish, then that is what we are desirous of learning.

To start aright in this search for the truth we must initially recognize that military organizations as such have no divine right but are suffered by a people simply because they have been able to give to that people security in a world where brotherly love exists more on the tongues of orators than in the daily practices of men and nations. In other words, national security is the aim; there is no obligation on the part of governments to provide this security in any certain way. There is no obligation to maintain an army, a navy, or an air force, if security can be provided without any or all of them.

So remember, we are searching for the truth and we are not asking *you* to believe simply because you think *we* believe. We hope common sense will be our key-word. We know that we are attempting to peer down the path of future warfare. We are not discussing the past; "Let the dead past bury its dead."[4] "Forward lie the pains of growth—backward the agonies of decomposition; to stand still is to rot."[5] Therefore, let us get on with our inquiry.

The first question we wish to answer for ourselves is "What is war?" As we study the writings of military leaders we come to the conclusion "War is a continuance, by violence, of a nation's peace-time policies."[6] You may remember that, several months ago, you were told that national policy was the strategy pursued by nations to insure security, prosperity, and ethnic unity. That, actually, is merely another way of saying that national policy is the strategy pursued by nations to insure their people an enjoyable self-preservation. Individuals want security and they want to be prosperous, and, since nations are merely collections of individuals and since national security and national prosperity mean an enjoyable self-preservation, it is readily apparent that nations, even as individuals, will stake all to obtain them. Self-preservation is the strongest of man's instincts; it is the same with nations. In the final analysis, all national policies are centered around the focal point of self-preservation. And self-preservation means prosperity and security; therefore national policy centers around prosperity and security. War is sometimes the method resorted to by nations to insure those two conditions. Therefore, war is truly a continuance, by violence, of a nation's peace-time policies.

Our second question—"Why does war occur?" Now in all walks of life, in all businesses, there exists competition. Great business establishments attempt to increase their size and their incomes by taking away from others; individuals attempt to improve their status, and this is usually done at the expense of someone else. In other words, this desire to change the status quo results in strife. Everyone wants to improve his position—no one wants to lose what he has—and, therefore, we find nations competing with one another in the markets of the world, in all of the activities which make them more prosperous, and, since this can usually be accomplished only at the expense of some other nation we see the commencement of strife.

Now nations try to adjust these differences by peaceful means—that is, in harmony with the instinct of self-preservation, for violence threatens life. However, frequently the time arrives when nations are unable to secure the things they consider essential for their continued self-preservation by peaceful means and they resort to war because war is the only other means available. There is an anomaly: first, self-preservation dictates to man that he remain at peace with his fellow men, and second, self-preservation dictates to man that he wage war against his fellow men. Peace and war are just two different means of securing the same end. Therefore, in answer to our second question we might well say with little risk of contradiction that major wars occur because the peace-time machinery has failed to enable a nation to continue those policies which it considers essential for the security and prosperity of its nationals.

Our next question—"What is the object of war?" Obviously, if war is a continuance, by violence, of a nation's policies and if war is resorted to only when the peace-time machinery has failed to insure a continuance of those policies, then the purpose of war is to compel an adversary to accept such policies. In other words, the real object of war is to overcome the hostile will. Major wars are not resorted to until some obstacle appears in the path of a nation and prevents it from carrying out its policies. That obstacle is always the policy of another nation which the first nation feels will, either now or ultimately, threaten its security or prosperity. That obstacle must be removed. If it cannot be removed by peaceful acquiescence of the nation imposing the obstacle, the only other remedy known is to apply force to the extent necessary to overcome that nation's will,—to make its will disintegrate to the extent that it offers no further resistance to the removal of that obstacle. When that is accomplished, the object of war has been attained. Therefore, the basic purpose, the fundamental object of war, is to force the will of one nation upon another nation; to overcome the hostile will.

The next question—"How has war been waged in the past and why has it been waged in that manner?" Since the object of war is to overcome the hostile will—to compel a change of policies to agree with the policies of the aggressor nation, then it should be apparent that wars have been waged only through the application of pressure, in some form or other, to a sufficient degree to disintegrate that will. In the past, when governments have been unable to accomplish by peaceful means those acts which they considered essential for the welfare of their nation, then they utilized the forces available to them to apply sufficient pressure to compel acceptance of their policies. This was accomplished through military forces. The aggressor directed its armies against the hostile nation. It moved its armies to occupy vital territory in order that, through the control of those areas, pressure could be applied to compel the hostile nation to capitulate. For this reason have we seen the army of a nation moving across a hostile frontier towards its objective: the area it desired to occupy. The movement of that army naturally met opposition because the other nation had also mobilized its military forces and thrown them across the path of the invader. A battle, or a series of battles, resulted, or, as in the World War, a gigantic siege took place. If the aggressor nation succeeded in defeating the defending army then it was possible to move, without further opposition, into the adversary's territory, occupy it, and exert the pressure necessary to break the hostile will.

Therefore, we find, throughout the pages of military writings, statements that the objective of a nation at war is the destruction on the field of battle of the enemy's main forces. Such a conclusion is a confusion of the means with the end. *The destruction of the military forces of the enemy is not and never has been the objective of war; it has been merely a means to an end,—merely the removal of an obstacle which lay in the path of overcoming the will to resist.* It was necessary, of course, that the hostile army be defeated, because so long as that army was free to act it could interpose itself in the path of the invading army and prevent occupation and the resultant application of pressure. However, the defeat of the hostile army was not the end sought; it was only a means to the end. *The end was the breaking of the hostile will.* Surely the truth of this statement is apparent if we could permit ourselves to visualize two armies meeting at the frontier and one army decisively defeating the other and then remaining where it is. Would the defeat of that army, without more, cause its nation to give in to the demands of the aggressor nation? Surely not; the victorious army must move forward and occupy vital areas in order to set up a condition which would be more objectionable to the other nation than the terms which the victor demanded.

Furthermore, even occupation, without more, might not cause the capitulation of the adversary. Something else must be done, and that something else is the application of pressure in some form or other. Again, if you will permit yourselves to visualize an army moving into another country and not interfering in any way with the functioning of that government, not exacting any tribute from its people, but, on the other hand, spending its wages in their shops and markets, buying its clothes in their stores and generally deporting itself in accordance with the customs and laws of that land;—would that constitute pressure? Quite the contrary; it would be a boon to the nation occupied. It is obvious, therefore, that occupation is not the end sought; that occupation merely places an army in a position to do those things which are so objectionable to the occupied nation that it prefers to accept the demands of the victor rather than continue in the state of occupation.

Therefore, we see that the defeat of the hostile army is not the end sought but only a means to the end. We see that the occupation of territory is not the end sought but only the means to the end. We see that the end is overcoming the hostile will to resist and that everything else is only a means to bring about this end.

As with armies, so it has been with naval forces. Navies have considered their objective to be the hostile navy. Why? Because through the destruction of the hostile navy it has been possible to act with freedom against the sea lanes of communication and deny to the adversary those raw materials and essential foodstuffs and the ability to move its military forces, all of which were required to continue to wage war and to continue to exist. However, the defeat of the hostile navy, without further action, would accomplish little. Let us suppose, for instance, that the naval forces of a coalition of powers defeated the British Fleet and then returned to its bases. Would that cause the capitulation of the British Empire? Of course not. It would then be necessary for the victorious naval forces to blockade the British Island and deny the movement of foodstuffs into England. Then pressure, the pressure of starvation, would begin to be felt, and finally capitulation and agreement to the demands of the victor would be more acceptable than to continue to resist without food on the table.

The reason why I have spent considerable time on this point is to make it clear that armies and navies have been the instruments utilized by nations for the sole reason that there existed no other means by which pressure could be applied. When peaceful means failed to bring agreement to one nation's demands upon another and it became necessary to create pressure, armies and navies were the only available means of making it possible to apply that pressure. Of course we all know that usually the defeat of a defending army was sufficient without occupation, but that was simply because the defeated nation knew that occupation was only a matter of time and that with occupation would come undesirable pressure. However, it has been the realization that occupation and pressure were possible that has caused capitulation.

Let me give one illustration to prove that the defeat of the hostile army is not always essential to bring about victory. In the World War, was the German Army, as such, defeated? Most certainly it was not. It could have retired beyond the Rhine and there remained undefeated, perhaps for years. Was the German Navy defeated? It most certainly was not. It was a stronger navy at the end of the war than it was at the beginning. As a matter of fact, through its submarines it almost succeeded in denying foodstuffs to the British Isles to the extent that Britain herself would have been seriously confronted with the spectre of starvation. What kind of pressure was it then which caused the capitulation of Germany? A study of the history of the Great War reveals that Germany was actually in a state of siege for four years; that, with its sea lanes closed, with access to the rest of the world denied to it, Germany was confronted with slow starvation. The morale of its people finally broke when they could see no hope of victory for its military forces. It was not the defeat of its military forces which broke the will to resist; it was the continuance of a condition, without any hope of change, which finally became more objectionable than the obnoxious terms demanded by the Allies. The German Army was not defeated or destroyed. The German Navy was still in existence. It was the breakdown of the German people through the continuous denial of those things which are essential, not only for the prosecution of war but to sustain life itself, that overcame the will to resist of the German nation.

Is it the advantage of civilization to change the past method of waging war if such a change is possible? The World War was waged by the traditional military forces in the traditional military manner. It was a war which has cost to date approximately two hundred billion dollars of the world's wealth; a war which snuffed out the lives of ten million combatants. It was a war which was nearly as costly to the victors as to the vanquished; a war which upset the economic equilibrium of the world for nearly a score of years, and resulted in the overthrow of the governments of many nations. In that war the prevailing idea of all parties was destruction; and in following this idea they nearly succeeded in destroying themselves.

Modern inventions—machine-guns, rapid-fire cannon, gas, etc.—have served to strengthen the means of defensive warfare and have made more difficult the waging of offensive war. Surely another similar world conflict might mean a breakdown of civilization itself. It would be a war of diabolic attrition, and in the end victory would go to the side whose internal organization, whose industrial organization, whose very social system,—though undoubtedly disintegrating,—had done so less rapidly than those of the other side. But of what profit would be such a victory? The vanquished side surely could not pay adequate reparations, because it would be like a sponge from which the last drop of water has been squeezed. With the delicate and intricate economic system of the world of today there can be no prosperity while the major nations of the world are suffering from economic paralysis; and so the objective of peace would be to restore the national health of all the belligerents—to repair the ravages of the disease: war.

Certainly it is not difficult for intelligent people to appreciate that the more speedily a war is over and the world can revert to its normal peace-time pursuits, the better it is for the entire world. Common sense dictates that every means should be employed which will bring wars between major powers to a conclusion within the shortest possible time and thus lessen the effect of the destruction of life and property and the disruption of the economic system. There is no other course if the preservation of civilization is worthwhile.

Surely, we can conclude that it is to the advantage of civilization and in the interest of the prosperity and happiness of nations to change the past method of waging war if such a change is possible. Whether future wars will be like those of the past, i.e., wars of absolute destruction, only the future will tell. However, it looks as though science has given to man a method of waging war differently.

Has modern civilization reduced or increased the vulnerability of nations? I have tried in the limited period to show that the real objective of war is to force the will of one nation upon that of another,—to break down the hostile will to resist.

There is plenty of indication that modern nations are interdependent not so much for the essentials of life as for those "non-essentials" needed to conduct their daily lives under the existing standards of living. When retrenchment in consumption and production in the last depression became manifest, it was soon world-wide—it was only a matter of time until the monster's tentacles reached into every so-called modern nation.

The trend in modern nations has been towards specialization in industry and agriculture, which makes for large territories which are not self-supporting. The miner has little to do with raising his own food nor with making his own clothes or other essentials for modern life. The city dweller is dependent upon other communities for nearly everything he consumes, and oftentimes the consumer and producer are brought together only through the medium of intricate systems of modern transportation. In large cities many of the workers are not self-sufficient even for their means of locomotion between home and work. Many factories and most homes and places of business are entirely dependent for electrical power upon sources far removed from their location. In some instance whole industries are dependent upon a single source of supply of raw material which must flow through a single vulnerable bottleneck;—witness the American steel industry and the Soo Canal.

Hence, modern industrial nations are much more vulnerable, because of the existence of the economic structure which our present civilization has created, than were the nations of a century ago, when the dependence of one section upon many others did not exist. It appears that nations are susceptible to defeat by the interruption of this economic web. It is possible that the moral collapse brought about by the break-up of this closely knit web would be sufficient; but connected therewith is the industrial fabric which is absolutely essential for modern war. To continue a war which is hopeless is worse than an undesirable peace, because the latter comes soon or late anyway; but to continue a modern war without machinery is impossible.

That nations today are more vulnerable than was the case a century ago, I believe, is obvious. This does not mean that nations can be attacked more easily; it means merely that they possess more susceptible features a successful attack on which would cause defeat.

Have science and invention provided a new method of waging war?

That is a question which you will have to decide for yourselves as this course progresses. We know that modern airplanes can carry bombs capable of tremendous destruction. We know that airplanes are being constructed with a range of thousands of miles. We know that one by one the limitations of aircraft are being removed; that blind flying and blind landing are now possibilities; that the danger of icing during flight has been eliminated; that navigation in the air is as exact a science as is navigation on the ocean. It is becoming more and more difficult for the scoffer of air power to find his famous "limitations."

What is possible for a modern air force, a real air force, consisting of a 1,000, or 2,000, bombers if employed against a nation's economic structure, is a question which we cannot call upon history to answer. We can, however, by using the common sense with which the Creator has endowed us, and by aiding that with a reasonable research into the dependence of modern civilization upon its present economic set-up, come to intelligent conclusions. We can determine, in our own minds, whether ground forces must undergo a radical change in organization if they are to continue to wage war in the future. We can determine, in our own minds, whether the British Fleet would today be safe in the inland sea of Scapa Flow, protected though it may be from submarines if England were at war with Germany or France.

We know that mountains, oceans, and deserts provide no obstacle to travel through the air. We know that barbed wire and trenches cannot be erected to offer obstacles to aircraft. We know that aircraft can fly by night as well as by day and we know that a ton of TNT does not lose its identity and become impotent merely because it has been dropped from the air.

There is one thing certain: Air power has given to the world a means whereby the heart of a nation can be attacked at once without first having to wage an exhausting war at that nation's frontiers.

Whether air power can, by and of itself, accomplish the whole object of war is certainly an academic question; but that the air phase of a future war between major powers will be the decisive phase seems to be accepted as more and more plausible as each year passes.

In conclusion, let me say that "philosophy is the love of wisdom and wisdom is the power of forming the fittest judgments from whatever premises are under consideration. If the aim of wisdom is to arrive at the fittest judgments, then indeed is common sense the true philosophy of life. In the great masters, common sense is not only spontaneous but prescient, for not only are actions adapted to circumstances, but the circumstances themselves are seen in advance of their happening."[7] That is what we are attempting to do,— to see circumstances in advance of their happening; for to be forewarned is to be forearmed.

Common sense must be our keynote and the conclusions we draw will better fit us to take part in the next war, if and whenever it comes; for it is the next war with which we are concerned—not the last one which has ended. Let the historians revel in that one—our concern is with the future.

Comments on "An Inquiry into the Subject 'War'"

George's introductory lecture to the Air Force Course provides the thesis statement for the Air Corps Tactical School's theory of strategic bombing: air power can win wars quickly, decisively, and independently. The originality of the theory lies in how victory will be achieved: not by the defeat of the enemy's military forces or by the bombing of its civilian population, but by disrupting the economic system of the enemy nation such that it will not only be unable to support its population but also be unable to produce the machinery required to continue the war.

Three key assumptions underlie the ACTS theory of victory. First, air power can attack the heart of the enemy without the need to first wage a war of attrition. George ignores the possibility of an attritional battle for air superiority having to be waged prior to being able to sustain deep-strike operations. Second is the assumption that a modern national economy is neither resilient nor adaptive enough to withstand a discreet number of air strikes against its vital nodes and/or chokepoints; such a course of attack will produce economic paralysis. Third, George presumes that economic paralysis will end a war quickly by collapsing the enemy's morale through economic deprivation and/or the inability to supply its military. The validity of these assumptions would be challenged in World War II. As it turned out, air forces were no different than surface forces in that the neutralization or defeat of the enemy forces remained a critical aspect of warfare. Further, modern industrial economies proved more resilient and required far more destruction than anticipated. Even when economic paralysis did occur, this did not bring about the immediate surrender of the enemy, a reality encountered with both Germany and Japan.

The three lectures that follow in chapter 2 build upon George's theoretical framework by considering further the offensive nature of air power, its ability to avoid conventional attritional battle, and its potential to achieve victory by targeting the center of gravity of the enemy—that is, the civilian population and its will to resist—by disrupting economic infrastructure.

2

The Objective of Air Warfare

This chapter presents three lectures that expand upon the ACTS thesis provided in "An Inquiry into the Subject 'War'" in chapter 1, asserting that the independent employment of air power had ushered in a new means to wage war. Foregoing the need to first defeat the enemy's military forces, air power could directly attack the enemy's will to resist by targeting the vital and vulnerable elements of its economic infrastructure.

AIR POWER AND AIR WARFARE

Major Muir Fairchild introduced the Air Force Course in 1939 with a lecture on the purpose of air power.[1] In "Air Power and Air Warfare," he emphasized the importance of having an air force in being, capable of achieving independent outcomes during times of both war and peace. He argued that it is the body of ideas and concepts developed during peacetime that determined the air force available at the outbreak of war.

Air Power and Air Warfare
Major Muir S. Fairchild, A.C.
March 27, 1939

Gentlemen, this morning we begin the main part of the Air Force Course.[2] You may perhaps remember that I outlined this part of the course in the orientation talk at the start of the short course on Basic Tactical Functions at the beginning of the school year. As was stated at that time, this main part of the Air Force Course will be divided into three general sections: Air Warfare, Air Forces of the Army, and the Employment of Combat Aviation. During this first section on Air Warfare we will attempt to develop the basic principles

Muir Fairchild, circa 1935.

of Air Warfare and the Application of Air Power. In the following sections of the course we will try and develop the status of the air forces of our army and consider their employment in various situations that might confront us.

In the short course on Basic Tactical Functions we tried to give you the coordinating principles governing the tactical functioning of Combat Aviation, as they have been developed by the School. Thereafter, the School, through the other sections of the Department of Air Tactics and Strategy, has laid before you the best and most up-to-date thought available on how the components of an air force operate to do their assigned jobs.

So far, however, little or nothing has been said about what we are going to use the air force for—how we are going to employ it—what the assigned jobs of its components should be. We propose now to set about remedying that omission. We will be concerned from now on *with what we want to accomplish* rather than how the different classes of Combat Aviation go about doing the jobs that are assigned to them.

The questions, then, that we want to try and find answers for, are: first, the big question as to what the mission—the purpose—of the air force is to be, and then, the subsidiary questions as to what the specific objectives are to be, that will best and most efficiently accomplish that purpose.

In your problems on ground forces you have been confronted with the typical decisions that are required of a ground force commander. They might be summed up in a single question, "Shall I attack or defend and where, when and in what manner?" The choices are relatively limited and the particular situation usually restricts the choice to only a few practicable alternatives. Furthermore, historical precedent in land warfare furnishes a sound basis upon which rather definite principles have been established. Even so, as I have no doubt you have discovered, the decision of the ground force commander is not always an easy one to make.

How difficult then, may be the decision facing the air force commander, with the widest possible choice of objectives and a practically non-existent historical background! There is no doubt that the likelihood for erroneous or faulty employment of air forces is very much greater than for wrong employment of ground forces.

For example, our latest bomber has a radius of action of some 2,000 miles. It is, therefore, theoretically possible for it to be employed against any objective located in the 12.5 million square miles of territory surrounding its base. This figure is, of course, meaningless in itself but it *is* an indication of the tremendous latitude that exists in the choice of objectives for the employment of an air force.

The lack of well established principles, developed from past experience, to guide the air force commander in directing the employment of his force, becomes all the more serious as the capacity of the force and the range of choice as to its employment, increases. In the ground arms courses you were given many rather definite principles to go on and historical examples of their application were, or could have been, advanced to support them. After a great war in which air power plays a major part it will probably be quite possible to give a similar set of principles covering the conduct of Air Warfare. At this time, however, it seems necessary to examine the whole subject of the application of air power in order to provide a proper background from which we may try and deduce the correct principles of employment, and the correct choice of objectives, for our air forces in the situations with which we may be confronted.

I should like to stress the great importance of this study—the importance of reaching a correct strategical conception of the employment of air power in general and, of course, of our own air power in particular. It is obvious that the aircraft that we design and build are constructed in accordance with our ideas of what their proper employment should be. Therefore, the powers and limitations of our air force are determined in advance, by our conceptions of how it should be employed.

The Objective of Air Warfare 49

You all know how long it takes to produce modern aircraft even if redesign to meet new requirements is unnecessary. The characteristics of the existing or M-Day air force cannot be materially altered within a year—if not longer.

It is very apparent that if we have had the wrong strategical conception of employment, and have developed an air force which cannot cope on equal terms with an enemy having a sounder strategical conception of employment—the results may be disastrous. Therefore, let us recognize the fact that developing a sound strategical concept of employment for our air forces is not alone a matter of importance for future Air Force commanders, on and after a vague and nebulous M-Day, it is a matter of pressing, immediate peacetime importance for every Air Corps officer today.

As an example of what an erroneous conception of the strategic employment of a new means of action in war can mean, consider the effect on Great Britain of the German submarine campaign of the World War. Before the beginning of the World War the submarine was evaluated as being one of the minor means of naval action, and the full results of its use were altogether unforeseen. As a result the British were immediately forced by German submarine activity to move their main fleet north to new, and previously unprepared, bases. The British authorities have since stated that the declaration by Germany in 1917 of unlimited submarine warfare brought Great Britain within measurable distance of complete exhaustion of her food supplies and hence of complete defeat. In other words they finally almost lost the war, because their inadequate strategic conception delayed until almost too late the development of counter-measures to submarine attacks upon seaborne commerce.

There is perhaps another lesson that might be drawn from this example. At the start of the war the Germans had only some twenty-seven submarines. Had they possessed several times this number with which to inaugurate the campaign there seems to be little doubt, in the light of results achieved, that it would have been successful. As it was, before they could accomplish their purpose, sufficiently effective counter-measures were developed to prevent accomplishment of that purpose. Let us remember then that not only must we have a proper strategic conception of how our means is to be employed, but we must see that the force provided has the capacity to accomplish our purpose. If we have not the capacity to accomplish what we set out to do—it will be better to accept another and less desirable mission until our capacity is equal to the task.

If we are to profit by the lessons of the past and prepare ourselves in advance to make the best and most efficient use of this new means of waging war, it is apparent that first of all it is necessary to examine the whole question of Air Power and its application.

Now Air Power we have defined as the immediate ability of a nation to wage Air Warfare. It is rather like naval power in this respect, in that it is Air Power in being, and not potential Air Power, which is to be considered. We are all aware that no matter what the potential capacity of a nation to create naval power may be, it is the fleet in being that is decisive. The same is true of Air Power. There is no possibility of throwing out covering forces and fighting delaying actions in Air Warfare while we prepare behind this defensive screen.

If an enemy has the power to strike us in Air Warfare, without effective counter measures on our part, he has the immediate power of preventing us from *preparing* to oppose him, in addition to taking other effective action against us. It is apparent therefore that Air Power to be effective must be in being at the time it is required and that we cannot count with safety upon improvising or mobilizing it after the need for its use has arisen.

We might, in passing, consider what the strength of the Air Power of major nations may be. Historically it is interesting to note that during the course of the World War, Great Britain alone produced a total of some 57,000 airplanes, and that at the end of war, she had some 22,000 in commission. While these were small light airplanes, not to be compared with modern aircraft in size or complexity, it is apparent that the Air Power of a first-class nation may be very great indeed compared to what we have been accustomed to thinking of during the past few years as a sizable Air Force.

I have here a chart showing a summarization of the strength of the German Air Force as of last June. These figures are from a confidential G-2 bulletin and should be treated as confidential.

It is important to note that all of these airplanes have been produced since 1935 and that they constitute a completely modern up-to-date air force with performance comparable to those of equipment we are buying this year.

Please note the production figures on the chart. In the case of Bombardment, for example, this will have resulted by the first of this month in adding some 2,900 additional new bombardment airplanes to the 3,350 shown. This gives a total of at least 250 active bombardment squadrons plus a reserve of about 50% of the 6,200 bombardment airplanes on hand. This bombardment

Table 2.1. Strength of German Air Force, June 30, 1938			
Type	Squadrons	Airplanes	Production Per Day
Bombardment	126	3,353	12
Light Bombardment	24	400	1
Pursuit	54	1,700	6
Patrol–Torpedo	8	247	
Observation	34	700	
Transport		1,500	17.5
Training & AMD		2,000	
Totals	246	9,900	36.5

strength of over 6,000 modern bombardment airplanes as the major striking component of the German air force, gives a graphic illustration of what we mean when we talk about *Air Power.*

It is necessary for any one who still may be thinking of Air Power in terms of a few observation squadrons, protected by some local pursuit and backed up by a scattering of bombardment and attack airplanes to radically alter his conceptions in the face of the Air Power that has actually been brought into existence in the world today. We must not be misled as to the potentialities of modern Air Power by the present size of our own GHQ Air Force.

Having in mind the facts and figures presented to you during the Bombardment Course, you might attempt to visualize the potentialities of such a force when employed directly against a nation as well as against its armed forces of whatever type.

Now Air Power, such as we have been discussing, is capable of waging Air Warfare. By Air Warfare we mean air operations in which primary reliance for the accomplishment of a broad purpose is placed upon the independent employment of Air Power. This broad purpose does not necessarily mean winning the war all by itself. It may well be to further the ultimate strategic success of the surface forces. However, the concept of Air Warfare does not include air operations in direct assistance to the immediate tactical success of particular forces. Such air operations constitute operations in immediate support, and will be treated under the heading later. Lest you get the impression at this point that this breakdown of functions is an invention of the School, let me assure you that it is based strictly upon War Department doctrine for the employment of our own GHQ Air Force, as laid down in TR 440-15.

You are, of course, already generally familiar with the historical background of air operations during the World War. These operations in the main constituted air operations in immediate support. We haven't the time to go over this ground. I *would* like, however, to speak briefly about the German efforts toward *Air Warfare* because this is the only complete historical background we have and it will be interesting to refer to some conclusions to be drawn from these efforts later on in the course.

The first effort in Air Warfare was the German airship raids on England. These were begun in January 1915, and during that year twenty attacks were made, in the largest of which five airships started and four actually got over England. Thirty-nine tons of bombs were dropped, most of which fell in open fields, apparently because the airship commanders either could not determine where they were, or were unable to bomb with any accuracy. Let us note in passing that a formation of ten B-17's would carry in a single mission all of these bombs that were dropped on England during the first year of Air Warfare.

During 1916 twenty-two attacks were made, in the largest of which sixteen airships started and fourteen got over England. Bombing results were not much better. No objectives of any vital importance were hit. About 130 tons of bombs altogether were dropped, *or considerably less than the capacity of one group mission of B-17's.*

By the end of 1916 the airship had been virtually defeated by the defense organization, on account of its extreme vulnerability. Throughout, the attacks were completely ineffective as to actual destruction of vital objectives. Some 500 people were killed and about 1,200 injured. Property damage was confined largely to small houses destroyed, particularly in London.

That sounds like a pretty sorry record. But I would like to quote to you some passages from the British Official History of the War, Volume III.

> The operations of the German airships over this country have been set down in some detail as a matter of historical record. The reader cannot fail to have been struck by the comparative ineffectiveness of the attacks, but it would be misleading to lay stress on the direct results. It should be remembered that the primary duty of the Zep-

pelins, and one which they splendidly fulfilled, was reconnaissance work over the North Sea, and that the raids on Great Britain used up only a minor part of their time and energies. Yet had the Zeppelins been built and maintained solely for the raids, it must be admitted that, from a purely military standpoint, they would more than have justified the money and ingenuity that went to their building. The threat of their raiding potentialities compelled us to set up at home a formidable organization which diverted men, guns, and aeroplanes from more important theatres of war. By the end of 1916 there were specifically retained in Great Britain for home anti-aircraft defence 17,341 officers and men. There were twelve Royal Flying Corps squadrons, comprising approximately 200 officers, 2,000 men, and 110 aeroplanes. The anti-aircraft guns and searchlights were served by 12,000 officers and men who would have found a ready place, with continuous work, in France or other war theatres. There was an observer corps of officers and men, and, in addition, some part of the energies of the police force and of the personnel of the telephone, fire brigade, and ambulance services was diverted to home defence activities.

And it must be remembered that this considerable force had to be kept in a high state of readiness all over the country, that it was, as it were, on constant active service against an enemy who, enjoying the advantages of the initiative, came from his sheds to raid when he chose.

There were many other effects of the raids, real and powerful, but difficult to assess. When raids were in progress, or even threatened, vital war work was held up all over the country.

The air-raid menace, more, perhaps, than any other aspect of the war, was responsible for a temporary revolution in English social and general life. Night brought the unrelieved gloom of darkened streets and a brooding sense of danger. The reader, with the full facts before him, may reflect on the paucity of the means which brought this about. The air war was fought out on the Continent of Europe, and the bombing of Great Britain was episodic. It is not difficult to imagine circumstances in which we might have been called upon to meet the full force of Germany's air strength over this country.

Unless the reader ponders what that implies, the air raids will not be viewed in their proper perspective, nor will the potentialities of air attack be made clear.[3]

Please bear in mind that the British Official Historian is evaluating results achieved by a few Zeppelins—never more than fourteen in number—results which by our present standards would be considered not only ineffective but complete failure. You are familiar with the capacity and efficiency of modern bombardment and attack operations. Balancing these factors against this historical example you may draw the inevitable conclusions for yourselves.

Now we have devoted considerable time to the summarization of this historical example of Air Warfare, because it is the only one on which we have sufficient knowledge to evaluate *means employed* against *strategic results achieved*. I hope you will thoughtfully consider these factors: The smallness of the force employed; the relatively few attacks that were made; the relatively small tonnage of bombs that was dropped; the inaccuracy of the bombing, and, hence, the inconsiderable material damage caused to vital objectives. Contrast these *means* with the strategic results obtained: The lowered industrial output; the disorganization and confusion; and above all, the breaking down of the moral resistance of the nation attacked.

Ask yourselves, in the light of these conclusions, what must be the result of Air Warfare waged with such Air Power as Germany has today established—over 6,000 bombers equipped to operate in all ordinary weather; equipped and trained to navigate to any objective within a radius many times that available in the World War; equipped to carry, on a single mission, many times the tonnage of bombs employed in four years of Air Warfare; and finally, equipped and trained to drop these bombs with scientific accuracy on the selected objective!

It is apparent that properly employed, the results must be almost incalculable. It is also apparent that proper employment is not automatic and inevitable. It is to be arrived at only through a full understanding of all aspects of the capabilities and limitations of Air Power, and a thoroughgoing air estimate of the national situation.

Since the World War there have been two examples of the influence of Air Power in being on the course of international relations so outstanding that they must be considered. The first example is to be found in the Mediterranean situation of 1935. Lest you feel that the importance of this incident is over-rated, I will give you an evaluation from a strictly impartial source. I quote from a publication of the United States Naval War College entitled, *The Employment of Aviation in Naval Warfare:* "Therefore, aircraft based on shore, as compared with ship-based aircraft, must be considered as having a strength much greater than would be indicated by an evaluation of numbers and types alone."[4]

Thus we may expect that a nation with strong air power can create zones along its coasts and around its outlying fortified positions within which its shore-based aircraft are likely to have practically continuous freedom of action, and in which their activities in the fields of information, security and attack will exert an important influence upon nearby naval operations. Before undertaking operations near hostile coasts, a naval commander must therefore seriously consider this aircraft influence, and decide whether or not he can afford the losses that are certain to ensue. To continue for extended periods to operate, or even to base, within the reach of strong hostile aviation it would seem to be necessary either to render the enemy air bases untenable, or to establish a deep and very superior defense around the fleet, neither of which is very easy of accomplishment.

The growth of the influence of defensive air power upon hostile naval operations close to shore has been strikingly demonstrated in the recent history of the Mediterranean. Great Britain is distant a thousand miles from the western entrance to the sea, but for two hundred years has completely dominated it through her navy alone. Her positions at Gibraltar, Malta and Egypt, although far from home, have been satisfactory for the maintenance of her fleet. In continuation of this policy of naval domination, and apparently without an adequate appreciation of present conditions, last year she moved the major portion of her fleet to the Mediterranean in order to halt the Italian conquest of Ethiopia. Italy estimated the situation accurately and defied Britain. Two significant events at once occurred: first, the British moved their naval forces to Gibraltar and Suez; and second, they then called upon France to assist them if they were attacked by Italy. Even though Italy was weakened through having to send her expeditionary forces through the Suez Canal directly past a large part of the British navy, this much superior naval force was constrained to abdicate control of the Central Mediterranean. Without doubt, Italian submarines and motor torpedo craft menaced the British Fleet, but in the World War the German submarines were an even greater menace. There was one element of opposing strength, however, which the British apparently had under-estimated until the arrival of the crisis, and that was the Italian air power.

To be able to nullify this power, England required at least an equal air power in positions which she could defend and supply, and so located as to be within flying range of objectives whose destruction would seriously injure Italy. England had no such positions. Her aircraft carriers were too vulnerable, and had to be moved away with the rest of the fleet. Neither the harbors nor the terrain of Gibraltar and Malta are suitable for maintaining large numbers of aircraft. Air bases at Cyprus or Alexandria would have been cut off from direct supply via Gibraltar, and are out of easy reach of vulnerable Italian positions. Hence Britain's call for help to France, which has a well-based air establishment even stronger than the Italian. Great Britain did not need the French navy, but she sorely needed the French air force and its bases.[5]

Here, then, we have perhaps the first outstanding example of Air Power in being accomplishing the purpose for which it was created. Note also that this result was achieved through the ability of Air Power to wage Air Warfare and not as an auxiliary arm in immediate support operations. While actual operations were not undertaken, this fact does not lessen the importance of this first outstanding example of the expression of modern Air Power, for actual operations were not required to achieve the purpose desired.

The second outstanding example is still fresh in your minds—I refer of course to the Munich agreement of last fall. What was behind this episode which President Roosevelt has so aptly described as "peace by fear"?

We have here the astounding spectacle of three of the world's major powers abjectly surrendering to the imperious demand of a recently vanquished and still impoverished nation; surrendering all of those basic principles for which they had in past times proved themselves willing to fight and die; surrendering all of the dictates of self interest, present as well as future; surrendering their well established moral obligations, backed by solemn treaties and equally solemn promises and assurances to abide by those treaties.

Here was something more important that the mere sacrificing of Czechoslovakia to prevent war. What of the billions of francs spent by France since the World War in arming Czechoslovakia against this very day? What of the repudiation by France of the treaty which the French Premier and Foreign Minister had repeatedly, recently and categorically affirmed a determination to honor? What of the European balance of power which Great Britain had consistently for over a century sought to maintain as her one fixed point of foreign policy? What of British and French interests in Central and Eastern Europe? These were turned over to Hitler that he might construct a Mittel-Europa on a basis of German ascendency extending to the western threshold of the Soviet Union. And what of the solemn treaty obligations of that Soviet Union to uphold Czechoslovakia as a buffer against that very threat from the West? Indeed, what of the precedent to be established as to the sanctity of international contracts—not alone as moral obligations, but as the very foundation of international law and future international relations? No one could believe for a moment that these sacrifices, great as they were, would satisfy Hitler for long in spite of his statements. He had previously assured the World and especially Czechoslovakia, only a few months before as he was gobbling up Austria, that he had no further designs on other peoples in Europe!

No, here was something even more than "peace by fear." It was the sort of peace that demanded sacrifices such as could formerly only be imposed upon a defeated and vanquished people in whom hope was dead.

But why should this be? Russia possessed the largest army in the world with 1,300,000 men under arms, and France alone had three times as many trained reserves as Germany. Even little Czechoslovakia could put 35 divisions in the field against Germany's 140, and the few passes in the Sudeten Mountains had been rendered all but impregnable by her extensive fortifications which were only exceeded by the tremendous defenses of France's famed Maginot Line.

On the sea the disparity was as great. Great Britain's mighty fleet, even without the help of France, not only offered complete protection but dominated the naval situation as completely as the most timorous could desire.

With such an estimate of the situation at hand, how can we account for the fact that the British and French prime ministers were almost mobbed by cheering crowds in the streets of London and Paris when they returned from Munich? Daladier had sacrificed France's honor; he had surrendered the whole of the French position built up at such heavy cost for twenty years in Eastern and Central Europe; he had capitulated to a new and greater German Napoleon, and yet the French crowds cheered him frantically for his accomplishment. Why?

Only one element is lacking in our estimate of the situation to make this astounding, incredible picture understandable.

German Air Power. 3,350 bombers on hand on June 30th and 12 more every day since then. A new squadron per day!

Here then we have true air power. We have seen what it can do—or rather what millions of intelligently apprehensive people believe it can do. The question which remains to be answered is: How is such Air Power actually to be employed to bring its full pressure to bear if and when that is required?

We will proceed in our next few conferences to examine this subject of Air Warfare in an attempt to develop the proper strategic concept for the employment of air forces, the instrument through which Air Power is expressed. We have tried this morning to indicate something of the effect that may be expected from one method of employment and also to show that there are many methods of employment. The selection of the proper method to fit the particular situation constitutes the problem.

Are there any questions or comments? Thank you.

Comments on "Air Power and Air Warfare"

The concept that an air force in being was essential to winning the next war was not new, as airmen had long been making this argument to justify an independent air force.[6] Muir Fairchild went further than most air power theorists by considering how air power could deter or coerce an opponent in both war and peacetime. The cross-domain influence of Italy's Regia Aeronautica in deterring the British Royal Navy in the eastern Mediterranean is an excellent example of a time when the threat, rather than the actual use, of air power had a significant impact on the outcome of a war. While Fairchild may have exaggerated the overall significance the Luftwaffe played at Munich, the air threat to London and Paris did contribute to Prime Ministers Chamberlain and Daladier's decision to appease.[7] This argument for the value of air power in peace and war provided justification for the Air Corps dream of becoming an independent air force.

Principles of War

Lieutenant Colonel Donald Wilson followed Muir Fairchild to the ACTS podium for his lecture "Principles of War."[8] Wilson concurred with Harold

George that the objective of war was not victory over the enemy's military forces but rather the defeat of the enemy's will. Wilson took it a step further, however, by contending that the true purpose of war was for the enemy to concede to terms favorable to one's national policy. Defeat of enemy forces or the enemy's will was insufficient if it did not achieve the nation's political objectives for going to war. He asserted that air power was best suited for achieving these objectives through massed offensive action. He contended that there were no "immutable" principles of war; these, in fact, changed over time. Except for the objective, the other principles of war—including mass, economy of force, movement, surprise, security, simplicity, and cooperation—were concerned with the employment of military forces. Wilson criticized the study of unchanging principles for ignoring how the weapons and methods of warfare change over time. He instead argued that airmen were better served by considering three fundamental factors: the end desired, the means available, and the application of those means.

Principles of War
Lieutenant Colonel Donald Wilson

1939–1940

Gentlemen:

    This morning we continue our search to find the TRUTH.⁹

    In fact this whole series of conferences which we group under the heading of AIR WARFARE is devoted to that type of search.

    If we are to develop sound strategical conceptions for the employment of Air Power, we must know with complete honesty and impartiality just what the nature of future warfare is likely to become.

    To visualize the steady march of military progress, we must conduct our search with an open and unbiased mind.

    In yesterday's conference Major Fairchild began the treatment of these important basic considerations by introducing the subject of Air Power—the immediate ability of a nation to wage air warfare—that kind of warfare which we believe marks the beginning of a far-reaching fundamental change in the means and methods of waging war.

    Our next step must be to briefly examine those concepts of the nature and conduct of war which have been developed as a result of the study of successful and unsuccessful traditional methods of warfare. Surely if we can find

Donald Wilson, circa 1945.

the basic truths concerning war, we can learn them and can hope to avoid the mistakes of our predecessors. This is encouraging, and we dig into military writings with a feeling that we are on the right track. We therefore turn our attention first to an analysis of these methods of past warfare which have been passed down to us under the familiar heading—"The Principles of War."

"The Principles of War are immutable," our Training Regulations stated in 1921, after listing at that time *our* nine principles of war: The Objective; The Offensive; Mass; Economy of Force; Movement; Surprise; Simplicity; Security; and Cooperation.

How does a *principle* become immutable and just exactly how can a principle be determined in an art in which controlled experimentation is impossible? In the case of the principles of war, some evidence is available.

Major General J. F. C. Fuller, of the British Army, relates that in 1911 he became convinced that war might break out at any moment and so started to prepare himself for the inevitable struggle. He turned to British Field Service Regulations and found the statement: "The principles of war are neither very numerous nor in themselves very abstruse, but the application of them is difficult, and cannot be made subject to rules."[10]

Familiarity with the methods of "writing" regulations should cause us no surprise in finding that this statement seems to have been derived from Marshal Marmont's dictum that "general principles for the conduct of armies are not very numerous, but their application gives rise to a great variety of combinations, which it is impossible to foresee and to lay down as rules."[11]

Strangely enough, the simple principles mentioned were not to be found listed or discussed in the British regulations. So General Fuller turned to the correspondence of Napoleon and from his study deduced six. Napoleon, incidentally, always refused to make a statement of *his* principles. "If I were to write the principles of war," he said, "their simplicity would be astonishing."

When a new edition of British Field Service Regulations was published in 1920, General Fuller's principles appeared therein, with a slight modification. The same principles, with one additional, were copied into United States Army Training Regulations in 1921, and it was at this time that this particular set of principles became immutable. He has continued *his* studies and re-stated *his* principles.

But the immutable principles disappeared from our regulations in 1928. However, they still are to be found, now six in number, with corollaries, in current *British* Field Service Regulations. They also appear, seven in number, in an American official text on strategy, published in 1936, and strangely enough, this set was copied from British regulations of 1924.

Not the least of the peculiarities of the principles of war is that other armies have discovered entirely different sets. French regulations, for example, list but three: Impose your will on the enemy; conserve your liberty of action; and economy of force.

Foch wrote a book entitled *The Principles of War.* He listed four such principles: economy of forces; freedom of action; free disposal of forces; and security.

This analysis shows that our American military thinkers have been content to accept *their* principles second- or third-hand from foreign military writers.

This also brings to our attention that the term *Principle* was always used by the elder military writers to mean a *guide* or *rule* of conduct.

This term—*Principle*—is defined in half a dozen ways, but in military writing it has not been used in a scientific sense as meaning a *fundamental* or *basic truth,* until our own Training Regulations called our present principles of war *immutable* or perhaps in simpler terms "unchangeable."

Based on our own regulations and concepts, we must first make sure that we understand a "principle" to be a *basic* truth upon which *many* other truths depend and, in the scientific sense, a truth that fits all cases.

However, we must recognize it as an historical fact that soldiers have always sought a mystical "by-path" or a "never-failing" formula which they needed *only* to follow in blind *faith* in order to gain the road to victory.

Therefore in our search for *basic* truths we must be ever mindful that while many of the so-called "Principles," "Maxims," or "Guides" of the past may have been excellent guides of conduct for Napoleon or either great leaders, they are not necessarily so for us. Our problems differ from theirs. The formulae of yesterday cannot be used blindly today.

With this thought in mind let us examine our present accepted "Principles of War."

The Objective

First, the Principle of the Objective. This principle is closely associated with an important basic factor—"The End Desired."

But what is the "end desired" in war?

We go to our primary statement of military doctrine as published in the latest Tentative Field Service Regulations—1939, FM 100-5—"The ultimate objective of all military operations is the destruction of the enemy's armed forces by battle."

Let us try to reconcile this with the happenings in the latter part of 1918. We see that although the German army *might* have been the ultimate objective of the Allied forces it was not defeated by battle. But since Germany was reduced to such a point that the Allies were able to dictate the terms of peace; something other than defeat of the military forces must have happened! *We believe it was the collapse of the German nation as a unit* and this causes us to score a major error against the Allies for not recognizing the most vulnerable element of that unit—for bending every effort to reduce the "military front," while not knowing that the "home front" was the weak link. However, the leaders of that day are not so much to blame, as would be those future leaders who might repeat the same mistake now that they have a *weapon* which can reach that "home front."

As a result of our analysis we must decide that the real *objective* in war is *not* the destruction of the enemy's armed forces by battle but is simply the defeat of the enemy nation to that degree necessary to force the enemy to sue for peace on terms favorable to our national policy. The *method* to be employed in bringing about this desired end surely is incidental. Certainly it should not be allowed to assume the importance of a paramount Principle.

The Offensive

The Offensive is obviously a *method*—it is one of the ways in which a force may be employed.

At the outset of the war in 1914, French combat orders carried the injunction, "The enemy will be attacked wherever found." The world was treated to the incredible and pitiful spectacle of 329,000 Frenchmen killed in the opening two months of the war during which this Principle of the Offensive was blindly followed. They believed the offensive was necessary and profitable on all parts of all fronts. They suffered a severe jolt in learning that they were in no position to resume the offensive on any front. From this we learn that it is certainly unprofitable to take the offensive *everywhere,* only to be repulsed *everywhere.*

From the national viewpoint a major offensive is the appropriate form of action only for those *forces which are expected to bring about a major decision.* It does not follow automatically that all of the command nor all of the types of forces available must be used on the offensive. Moreover, in most cases it is essential to assume the *defensive* with a large proportion of the total effort in order to release the power needed for the *offensive.* This has been accepted, even by the French now, as reasonable employment for *ground forces*—it is yet to be applied to the several types of forces available on the national scale. This is evidenced by the fact that all types of national military forces, land, sea, and air forces, make the offensive their goal with little regard to their peculiar characteristics in this role.

We do claim that *offensive* action should be predominant with air force, but this is not because of blind adherence to a so-called "principle of the offensive." It is simply because we believe that offensive air action is decidedly stronger in producing the national end desired. We believe that the offensive capability of an air force is its *most outstanding characteristic.* We believe that intelligent employment must exploit those known characteristics.

The Principles of Mass—Economy of Force—Movement

Let us now consider *mass, economy of force, and movement.* Like the Offensive, each has its special application according to the type of force and the echelon of command under consideration. For example, the effect of *mass action* by an air force can be produced without physical grouping of the forces as is required in the mass of land and sea forces. Here again the characteristics of the force must be thoroughly appreciated and the application of these methods should be in accord with those characteristics and the end desired, rather than the blind following of a method originally devised for some other set of conditions. Even today you will find those who would employ the *different* characteristics of air forces as though these were the *same* as ground or naval forces.

Surprise

If there is any one factor that will produce large dividends it is this one—the Principle of Surprise. Surprise plays on the human characteristic of fear of the unknown. We all are likely to exaggerate the effect of anything which occurs suddenly and without warning. There are many examples in war where a very small force appearing at an unexpected place has caused havoc. A surprise *formation* defeated the Romans at Cannae. The surprise of the submarine almost defeated Great Britain in the last war. Apparently the surprise of air force action and its effect was a prime factor in the rapid defeat of Poland.

We should note from the Polish incident that surprise methods and weapons work very well on the national scale. Surprise is not confined to local engagements on the battlefield. However, we note that surprise is often neglected on the national scale, particularly by the "have" nations.

If we know definitely that our national structure would be attacked, say in ten or twenty years, and that our enemy would place his main dependence in air attack, we would now go about the job of rearranging that structure and its defenses to best meet such an attack. On the other hand, if we adhere to the belief that the war must be decided on the high seas or in the mud, and along comes a serious attack against our "home front," we may be sufficiently surprised, shocked, and damaged to be unmistakably defeated. Just as complete and effective surprises have been wrought in past warfare. Human reaction is the most permanent element in all history. It is the element upon which surprise is registered. Hence, the factor of surprise will always be a most powerful weapon.

Security

Security or protection is largely concerned with *preventing* surprises. It embraces the measures employed by a nation or a military commander to prevent the occurrence of a condition for which no adequate provision has been made.

The greatest difficulty with providing security is that it usually reduces the efficiency of other elements, particularly when the available force is limited. The proper balance between the need to guard against the effects of enemy action, while at the same time exerting pressure against him, is the proper solution for the factor of security.

In arriving at this compromise it is necessary to realize that the minimum security demanded is that which will permit continued operation of the force or activity until it can accomplish its purpose. Extreme measures for security, such as removing installations beyond the reach of the enemy, may nullify their purpose and hence may be entirely impracticable. At the other end of the scale, lack of security measures may cause quick defeat and thus prevent the force from striking a single blow as experienced by the Polish Air Force. Somewhere between these extremes is found the compromise which provides the proper balance between *effort expended in security* and that which *remains for use against the enemy.*

Simplicity and Cooperation, like nearly all of the other items we have considered, are factors concerning *method.* They are both important factors. Both are made necessary by human limitations.

Simplicity is demanded because of the lack of facility of the human mind in grasping and coordinating more than a very few considerations at one time. This lack of mental capacity is so universal, and so well recognized, that military men have raised the desire for simplicity to the exalted position of a "Principle of War." The chief point to remember is that effective execution demands a simple plan—even though a great amount of complicated, tedious, painstaking mental effort on the part of the commander and his staff may be required to produce the most effective simple plan. Too often such mental effort is avoided by the commander and his staff under the guise of simplicity. The Polish solution to the problem of defense against a modern military force was quite simple but certainly disastrous! Hence, we must conclude that simplicity is not an advantage except as it helps to produce the desired end.

Cooperation is essential because of the human incapacity to foresee all contingencies and provide the necessary measures for *coordination.* We all know that generous, willing, and voluntary cooperation is no problem when morale is high and everyone is driving toward a *common purpose.* But we must not forget that wholehearted cooperation throughout all ranks is essential to this end.

In continuing our examination with a view toward progress, we may at times appear to be somewhat critical. However, we want carefully to avoid being destructive—we are critical only to engender the sort of dissatisfaction which is the essential preliminary to progress.

If efficient warfare always had been the rule in the past there would of course be little concern for the future. But we know that this has not been the case—mostly because of the resistance to new weapons and new methods. At least this resistance has been the rule for those who had to be *shown*—those who lacked the *vision* and *courage* necessary to accept a promising theory. These—the conservatives—are always in the majority.

This majority opinion in the military is an average intelligence which resists the more progressive ideas while hanging on to those which have outlived their usefulness. Not many months ago it was dangerous to speak of airplanes in the thousands, simply because the majority could not grasp such an idea so long as it remained a theory. Now we scurry to secure thousands of airplanes and argue about what to do with them.

Opposed to this average level of thought and intelligence, there is always a small, enthusiastic, and troublesome minority.

This minority is often wrong, but sometimes it is right. Of course the amount of noise and trouble it creates is no indication of whether it is right or wrong. Fortunately though, when it is *wrong* its life is short; when it is *right* it persists despite all opposition and thereby leads the way to the future.

However, in military matters particularly, much *time* is required to change the mind of the majority. Even when the correctness of an idea has been demonstrated to the majority it is still frequently not accepted. Such demonstrations must be repeated *time* and *time* again before the conservatives are convinced.

For example, the double envelopment method of attack was demonstrated 490 years before Christ, when 10,000 Greeks decisively defeated *twice* as many Persians at the battle of Marathon.

This would appear to be a rather clear demonstration of the value of this method, under the conditions of that time—and to some it was—*but* to the conservative majority it was just another "special case." This is borne out by the fact that 274 years later, with all those years to absorb the lesson, it was learned, apparently, *only* by the small and troublesome minority and discarded by the conservative majority. This is no exaggeration because the majority was represented by the Romans—86,000 strong—who allowed the young upstart Hannibal with little more than half that number of men to repeat the tactics of Marathon and practically annihilate the Romans. *This was the famous battle of Cannae.*

Now this is not particularly important, except that the battle of Cannae repeatedly has been quoted by military students as a model for *annihilation of the hostile force* without so much as a word about the slow thinking on the part of the smug Romans.

As another example of slow learning, this time in connection with a new weapon rather than a method, we might cite certain European experience more than one thousand years after Cannae. In the year 1346—19,000 Englishmen demonstrated to more than *three times* that number of Frenchmen the value of the *longbow*, as opposed to mounted knights in armor.[12]

The English demonstrated this new weapon so well that half the French were lost while the English lost the ridiculously small number of fifty men.

This appears to be a rather clear demonstration of the effectiveness of a new weapon. Moreover it was not a battle in some far off corner of the globe—it occurred on French soil. Did the French learn their lesson? Was the new weapon something they could not produce? Were they given any other demonstration that might serve to confirm this experience?

Only ten years after the first demonstration, the longbow was again victorious. This time 6,000 English decisively defeated 16,000 Frenchmen. Certainly the lesson must be learned now! Let us see.

After the French had had seventy years to contemplate the adoption of the longbow, they marched 50,000 mounted knights and humble foot-soldiers into the withering fire of a mere handful of Englishmen whose principal weapon was the longbow (year 1415, near Calais).[13] Nearly half the French force was lost—but, had they not maintained their traditional weapons and methods? Apparently the French were saved from ever learning the lesson of the longbow because an even more revolutionary change in the form of gunpowder came into use.

These examples are dimmed by the obscurity of history—they appear to have little bearing on present-day activities. But we should reflect that these people and their soldiers, who clung steadfastly to their traditions and orthodox methods were seriously fighting to defend something they wanted to keep. To them, these incidents which we stamp as colossal blunders, were no more evident then as are the errors in more modern times.

Contrary to the popular belief, our modern times are not entirely illuminated with military brilliance.

For example, let's look at the French again. For at least forty years after their defeat by Germany in 1870, they had thoroughly schooled themselves in the belief that war consisted of no other method than the *offensive*. As we pointed out, that method had taken on a sort of religious fervor. When that war began in 1914 they plunged headlong into the offensive only to be repulsed with such losses as to permanently weaken their chance of success. Now certainly war was a serious matter with France, but still here was the whole thing based on a wrong concept—a failure to evaluate properly the factors—a blind following of a will-o-wisp misnamed a *Principle of War*.

But France was not the *only* offender. Germany was prepared for a few months' war and was unable to adjust herself to a contest which dragged into four years. England, with her habitual confidence in her fleet, failed to estimate the influence of the submarine and nearly starved before that influence was overcome. On the other hand, Germany failed to adopt wholeheartedly the more or less *new idea* of the submarine and found herself with about *one-half* as many as were required. This lack of the necessary number of submarines must be charged against the old line navy people in Germany who insisted upon building capital ships only to have them be bottled up most of the time and scuttled at the end of the war.

We ourselves failed miserably to grasp the meaning of modern war, even after watching it for two and a half years. We were pretty well prepared for another Spanish-American War but that kind of war was not being staged. *We entered the struggle in 1917 almost totally unprepared.*

Major General George A. Lynch, Chief of Infantry, in a lecture delivered to the class at Leavenworth made these comments:

> It is painful to have to recall that [although] the first three years of the World War had demonstrated the power of the machine gun, we not only entered the war without a machine gun but did not even have one under development—and this was not the fault of the supply services but of the tactical authorities who should have prescribed the characteristics of the weapon. It is likewise unpleasant to recall that

The Objective of Air Warfare 69

we entered the war with tactical regulations that called for the *deployment of infantry* at a density of one man per yard though the war had clearly demonstrated the stupendous folly of this procedure.[14]

And yet we were a world power entering upon the serious business of war.

Of course the natural reaction is that those things happened *then,* and that they cannot recur. To a large extent that is correct because most of those deficiencies revealed by the World War have been overcome.

Today there is no problem concerning machine guns; there is no failure to realize the need for more than one yard interval in extended order; and only recently we at last changed our infantry division organization to conform to that which was adopted by foreign armies during a war that took place more than twenty years ago. In other words we are now pretty well prepared to fight the World War over again.

But this does not mean that we will foresee the factors influencing the *next* war any better than these were foreseen in the past. It does not mean that we have grasped the significance of *"Air Power"* any better than we grasped the meaning of machine gun power. It does not mean that we cannot be surprised by some young upstart who is willing to proceed on the basis of a *promising theory* such as employing an air force composed of literally thousands of bombardment airplanes. Such a surprise was sprung at Munich, and again in a more definite manner in Poland.

It is obvious that mistakes have been made—it is obvious that traditional weapons and customary methods do not necessarily produce successful results. Hence, we must reach the conclusion that *military equipment and military methods must change* to meet the changing conditions. But where is the guide that will carry through this transition—that will show us things as they are—that will picture conditions in their true light? Certainly when we run on to an imposing title such as the *"Principles of War"* we are justified in expecting just such enlightenment.

But slowly, after much search for the basic truth concerning the preparation for war, and the conduct of war, we find that practically everything that has been written is devoted to an explanation of *methods* concerning the conduct of operations *on* or *near the battlefield*. Such explanation is based upon certain *definite weapons,* and only too often provides little or no insight into the many basic factors which are essential if we are to profit from that experience in any future situation.

Nowhere can we find any reference to a principle which teaches us to organize that *type of force* which will most efficiently produce the end desired. Though we do find in unmistakable terms—"Infantry is the basic arm and all other arms operate to assist in its success."

Here we pause and reflect on the noble knight. Undoubtedly he was the basic arm of the French! He *remained* the basic arm and the French continued to be defeated.

Now we must accept unmistakable signs of *changing weapons* and *changing methods*. Consequently, we are a bit doubtful about anything remaining a basic arm forever, and we can't be sure but that the change may well be made in the not too distant future.

Realizing that change *does* take place we are forced to believe that basic truths, immutable and unchangeable, cannot be based on any particular set of temporary conditions. All methods of warfare and all weapons of the past (as well as those of the present) are only a step in a continuing development.

Our analysis of the concepts developed from the study of past warfare has been none too fruitful in providing us with immutable principles upon which to base our conceptions of employment of Air Power in future wars. Having become sufficiently discouraged in looking for a *guide to the fundamentals* in the conduct of war, we give up and determine to accept these basic factors upon which we will predicate our future investigation into this subject.

1. The end desired.
2. The means available.
3. Application of the means.

The end desired. (This may be called: The Objective to be attained—the purpose—the aim—the object sought—or, any number of terms—but in the final analysis it is the primary purpose we choose to call "The end desired.")

The means available. In war on the national scale this embraces the whole of the nation's resources of every kind. (In *military operations* the means includes all of the various types of forces or other methods by which the command may exert pressure on the enemy to bring about the end desired.)

The application of the means. This is the action which is intended to bring about the end desired, by use of available tools or forces—*the means available*. Intelligent application of the tools available to any echelon of command presupposes the most advantageous use of the *peculiar powers* of each weapon or other means employed.

Well, we have been through the items that often pose as "principles." We see they are almost wholly items for consideration in the application of a force to produce a desired end.

As such, of course they are important in all echelons of command.

However, as our analysis has shown, to dignify these *guides to correct application of the means* by giving them the name of *Principles of War* and utilizing them as the basis of our strategic concept for the application of Air Power,—is more than likely to lead us to disaster.

This idea has been aptly expressed by a recent (and advanced) article published this year by an American Military writer:

> Military catchwords, such as the principles of war and quotations and maxims from the ancient great, have caused the loss of more battles, the deaths of more men, the stultification of more thinking, than treason, cowardice, or stupidity.
>
> It is evil to approach war with fixed ideas; that is, without an open and flexible mind, but it is certain to lead to disaster to approach it with the inapplicable formulas of the past.[15]

Gentlemen, approaching our study of the proper application of Air Power with open minds and with a complete disregard for the catchwords and slogans of the past, we will proceed in our next few conferences to investigate this matter of the proper application of Air Power in the light of these basic factors:

1. The end desired.
2. The means available.
3. The application of the means.

Are there any questions or comments?

Comments on "Principles of War"

Wilson rejected the conventional military view on teaching the principles of war as previously practiced at the school.[16] He instead argued that technology invalidated "immutable" principles and cited historical examples of how the English longbow in the 14th century, the machine gun and submarine in World War I and, most recently, German air power in Poland in 1939 had changed the nature of warfare. In the section on the principle of the offensive, however, Wilson paradoxically criticized the French in World War I for their obsession with the offensive, but then concluded that the offensive in fact remained the predominant characteristic of air power. In the end, the only backing for his claim was the school's corporate belief that this was so. Still, the lecture is an important contribution for its emphasis on connecting means to ends: how massed offensive air action could lead to desirable political outcomes.

THE AIM IN WAR

Captain Haywood Hansell's lecture "The Aim in War" expands upon Wilson's discussion on ends, ways, and means by specifying that war is a means by which a nation can secure its national interests of prosperity, security, and racial unity (in the case of Nazi Germany).[17] The objective of war is to overcome the adversary's will, a force that resides with the people. He argues that the population is the enemy's center of gravity, regardless of the type of government, whether democratic or authoritarian. As evidence, Hansell points to how World War I ended with the German Army undefeated and deployed on foreign soil, while the German monarchy was defeated back home once the people lost their will to fight.

Hansell further contends that a nation must also have the means to continue fighting, and that means and will are interrelated: a nation that loses the means of waging war quickly loses its will to do so and vice versa. In the past, to break the people's morale meant occupying, or threatening to occupy, their territory. To prevent such an outcome, nations raised armies and navies for defense. The defeat of these armed forces, that is, the means to continue war, is thus an intermediate objective to occupation. Hansell concludes that, though the desired end of war—the breakdown of the enemy's will—has not changed, the means to achieve that aim *has* changed now with the arrival of the airplane. Air forces, unlike surface forces, do not have to first defeat enemy armies or occupy territory. Such obstacles can be circumvented for an immediate and direct attack on the will of the enemy population.

Haywood Hansell, circa 1931.

The Aim in War
Haywood Hansell
September 9, 1936

During the war between Austria and Prussia in 1866, a German staff officer, Verdy Du Vernois, found himself confronted with a difficult situation.[18] He was a great student of military history and was familiar with the campaigns of all the great captains. Seeking to find a solution to his difficulties, he repaired to his tent and reflected upon the military operations of those illustrious leaders. Nowhere could he recall a situation—and its attendant solution—which seemed analogous to his present predicament. Returning to his associates, in disgust, he remarked, "To the devil with history and tradition; what is the problem?"

Gentlemen, we find ourselves today in a very similar predicament. We are approaching a study of the strategy of air warfare. Before we can arrive at an intelligent solution we must know the conditions of the problem.

Regardless of how familiar we may be with the methods and operations of great military leaders, we must understand our *peculiar* problem before we can derive the benefits of past experiences. If we are to profit by the military strategy of the past, then we must do so in light of conditions of the present.

Strategy has been defined as "the art of the general" (and that is, in fact, its literal meaning in its Greek derivative), and again, by General Von Molke as "the art of applying the means placed at a general's disposal to the attainment of the object in view."[19]

Those great military leaders, whose victories have been recorded in history, have been true artists in the sense of the latter definition. They have taken the means at hand and applied them in the most efficacious manner toward the achievement of the end desired. In many instances, the methods employed were, at the time of their employment, unorthodox and radical. However, two factors constituted the basic foundation which elevated those leaders from the ranks of skillful *artisans* into the select fold of great *artists*. First, they had a clear conception of the end desired; second, they were intimately familiar with the powers and limitations of the force at hand.

It is only by approaching our problem through those two fundamental stepping-stones that we can hope to gain an insight into the strategy of war in general, and the strategy of air war in particular.

As professional soldiers, we have an additional function and responsibility. We are concerned not only with the strategy of war. We are concerned with the strategy of peace as well. That is particularly true of military students of air warfare. The Air Force is, in very truth, in its infancy. It is our duty to mold and direct the development of this growing giant during time of peace so that he will be best fitted to accomplish the end desired of him in time of war.

Again we return to that first fundamental—the end desired. Surely, it is only through a clear conception of that ultimate aim—that end desired—that we can plan and develop our force for its maximum effort. With this object in view, let us analyze the ultimate aim in war.

We are, all of us, professional soldiers. We are being maintained, in time of peace, by a government for some eventual purpose. What is that purpose? We are an expensive luxury. What is expected of us? Why do governments maintain professional soldiers? *Why* is a soldier?

It is, of course, self-evident that we are being trained and equipped for war. What then is war? What is the aim in war? What is eventually expected of us?

To answer the first question—what is war—let us quote the eminent German author, Von Clausewitz. About a century ago he entered upon a searching analysis of war. He, too, found it necessary to establish, first of all, the purpose behind armed conflict between nations, and so he defined war as "the furtherance of national policy by other means." Let us note that war is, by no means, the *only* means. Von Clausewitz realized that there are normal means for furthering national policy in time of peace. Von Clausewitz had in mind diplomatic, economic and financial stratagems by which nations seek to further their own policies in time of peace. It is only when all other means have *failed* that the conflict is continued by violence. War is not an evidence of a generative conflict between nations; it is simply an evidence that all normal means have failed to settle an old dispute. War is a furtherance of national policy by violence. Since nations find the real fulfillment of their policies in peace, the real object of war is not a continuance of violence, but the establishment of a satisfactory peace.

From the standpoint of the soldier, the object of war is the restoration of peace on terms favorable to the national policy of his own people.

What is meant by this expression, "National Policy," which we have so often used? It is defined by Simonds as "the strategy employed by a people to obtain prosperity, security, and racial unity."[20]

If we accept that definition, and we further accept Von Clausewitz's definition of war, as the conflict between national policies, then it becomes apparent that war is, in general, the effort of people to find, through violence, their prosperity, their security, or their racial unity.

A glance at the map of Europe, after the Treaty of Versailles, seems to authenticate that concept.

Prosperity—Security—and Racial Unity. Let us consider the first: Prosperity. Prosperity is an expression of economics. The Great Powers today are all manufacturers. They find prosperity through the conversion of raw materials into manufactured articles. For that prosperity two factors are requisite: a source of raw materials and a market for finished products.

England today is an example of a well-balanced economic empire. Seventy million Englishmen govern and control four hundred million subjects. One-fifth of the world's land area is under British dominion. India alone is a tremendous source of raw materials and contains over three hundred million purchasers for British goods. That economic empire was carved through military conquest. It will have to be maintained by military force. England's policy is concerned with maintaining that advantage. Her problem is to keep the land which she has acquired, and to preserve the lines of communication which bring her raw materials and return British goods.

Across the channel 66 million Germans are crowded into a small corner of Europe, without any colonial empire whatever. The only source of raw materials for Germany is in her own backyard. She must find her markets by the hard process of competition. Germany's economic prosperity cannot be maintained under her present status. She must expand or face economic ruin. Her policy conflicts with the policies of those other nations which are wealthy and satisfied. If diplomatic and financial measures fail to solve her problem, then she must eventually take the only remaining course: she must fight for her national prosperity.

Again, Italy! Forty-four million Italians crowded into a barren peninsula, totally lacking all the essentials of economic prosperity. Italy lacked even the financial weapon for she was poverty stricken and faced with an annual and increasing deficit. She sought solution to her economic difficulties by a war of acquisition in Africa.

As for the second factor in National Policy: Security. France today, possessed of the world's second largest colonial empire, is desperately concerned for her security. Her policy is concerned with maintaining what she has.

We, ourselves, are concerned principally with security. We are concerned with keeping what we have acquired. Although we have less than 6 percent of the world's population, and only approximately 6 percent of its territory, we own and control:

40 percent of the world's gold
49 percent of the world's copper
52 percent of the world's lumber production
56 percent of the world's cotton
67 percent of the world's oil

And yet our internal prosperity is in large measure influenced by our foreign trade. That, too, we try to maintain in the face of competition. As an example of the conflict which rages in time of peace—consider the conflict of our policy with that of Japan in the Far East. Our policy demands the "Open Door" in China, as a market for our goods. Japanese policy is clearly directed toward commercial subjugation of China and the reservation of that vast potential market for Japanese exploitation. At present the conflict between American and Japanese policies is being waged essentially by diplomatic and economic pressure. Whether or not the problem can be settled by peaceful means remains to be seen.

As for the third factor: Racial Unity, we have only to glance at the map of Eastern Europe to find a maze of races grouped into arbitrary, and often unwelcome, nationalities. People of the same race are prompted by a fundamental urge to unite. They desire an autonomous government embracing all the people of their race. If they cannot find political freedom by peaceful means, they will fight for it if they are able to do so.

These desires to find prosperity, security, and racial unity are, of course, an inevitable source of conflict in a world already crowded and jealously exploited. Where national policies cross, a dispute arises. There is, unfortunately, no authoritative World Tribunal for the settlement of such disputes—no impartial court whose decisions can be enforced. Conflict of policies, if sufficiently vital, leads to war.

The essence of war is, of course, conflict. It is an attempt to settle a conflict of policies by a conflict of arms. It breaks out when further endurance of an unsatisfactory peace becomes intolerable—or when acquisition by force seems within reach.

Since war is a conflict, what are the fundamental factors underlying its expression? It is essentially and fundamentally a conflict of wills—the will to obtain is opposed by the will to retain. The will to progress is in conflict with the will to resist that progression.

We said that war is a furtherance of national policy by other means. It is a continuance of an old fight. It is an effort to decide, once and for all, whether the will to progress shall gain the ascendancy over the will to resist. Hence it is, in the ultimate analysis, an effort to overcome the will to resist. Where is that will to resist centered? How is it expressed?

It is centered in the mass of the people. It is expressed through political government. The will to resist, the will to fight, the will to progress, are all ultimately centered in the mass of the people—the civil mass—the people in the street. They represent the mainspring of every national machine. They are literally the heart of every national structure. Ultimately, defeat or victory is determined by the breakdown or the triumph of the citizen at home. If there can be any question of the truth of this statement, we have only to consider the last war for verification. The German war machine collapsed because the German political empire collapsed. Germany's armies—her military forces—were far from being defeated. They were, in fact, still on foreign soil and their Homeland was intact. Defeat was finally expressed because the people at home, the body politic, had lost its willingness to fight, because the will to resist of the German citizen had been broken. As in every other case, when to continue seemed hopeless, the cause was lost. That example seems particularly conclusive because of the type of government prevalent in Germany. Germany was an absolute monarchy, dominated by a war-like military machine. If there can be any doubt that ultimately the people are sovereign, that governments are accepted or tolerated by the people, then surely that doubt should be dispelled in the light of German experience.

The mainspring of national government lies in the mass of the people. Only so long as those people maintain their will to fight, can any government continue to fight. Whatever may be the morale in the armed forces, the loss of morale at home is decisive.

Gentlemen, in laying such stress upon the enemy's "will to fight—or will to resist," we presuppose his *ability* to fight. However willing he may be, obviously he must also have the *means*. When it comes to armed conflict, he must have *armed forces*.

Von Clausewitz has also defined armed forces for us. He said, "Armed forces are *instruments* for the furtherance of national policy."

A nation which pursues policies likely to provoke conflict, must have the means to enforce its desires and the will to employ those means. The two factors go hand in hand: the desire to do something, and the means to do it. If we are to frustrate the enemy's intention, we may accomplish our purpose in either of two ways: we may break down his will to fight, or we may break down his means to fight. The two are, of course, intimately related. Past experience has shown that the breakdown of the national will quickly follows the defeat of the armed forces, and conversely the political government whose will has been sapped finds military defeat insured.

Both these methods—these approaches to the ultimate end desired—are equally conclusive, and both have been used—separately or in conjunction—in past warfare. The siege of the walled city was an effort to starve the enemy into submission—and whereas both the means to fight (the military element) and the will to fight (the government represented by the people)—were affected, the primary burden fell upon the latter—the people. If the city capitulated without a final assault, it was because the people preferred the terms of peace imposed, to endurance of further hardship. The garrison might still be willing and able to fight—the means to fight might still be present—but the will to resist, as expressed through political government, had been broken down. As another example, the Central Powers in the last war found themselves in a virtual state of siege by the Allies. The means to fight still existed at the Armistice, but the seat of political government was no longer willing to fight. The people preferred the terms of peace to further endurance of the hardships of war. Many military experts are agreed that Germany's military forces, fighting on internal lines, might have resisted defeat for years. The war was finally decided by the breakdown of the national will to resist.

Hence, the ultimate aim of all military operations is to destroy the will of those people at home. This is, by no means, a new or radical doctrine. On the contrary, it has been accepted as the ultimate aim of armed forces for a long period of time. If that is so, then why have not military forces in the past proceeded directly to that goal? Why had not armies placed themselves in position to control that will of the enemy people to resist? That objective, too, has been recognized for a long time.

Every commander of land forces has recognized the necessity for ultimately breaking the will of the people of the hostile nation. The only means through which that objective could be achieved lay in occupation. The ground force exerts conclusive pressure by occupying territory vital to the enemy people or by the threat of such occupation. However, the characteristics of ground forces and ground warfare have imposed certain limitations on this process. The hostile nation, recognizing the vulnerability of its civil populace, has itself maintained armed forces to protect those vital elements. It has interposed its own armed forces to form a barrier between the aggressor and the heart of its own nation. Before the aggressor could reach that sensitive civil will he had to break through or to destroy that barrier. The aggressor had no choice in modern times. He could not go around the enemy armed force because he, too, drew his real vitality from the heart of his own nation; because he had to maintain the arteries through which that energizing flow of men and supplies reached him, from his own people. Any attempt to avoid the defending armed force and strike at the heart of the enemy nation, left exposed those vital arteries of supply. Hence, the ground commander was forced to accept an intermediate objective; he was left no choice; he must defeat the enemy's armed force preliminary to final military pressure through occupation. This concept was so widely recognized as to become confused in the minds of many with the ultimate aim. Many leaders came to accept the enemy's armed forces as the true military objective—as in truth it might have been since it was an unavoidable preliminary. Even Von Clausewitz himself inclined toward this error when he defined strategy as the art of the employment of *battles* as a means to gain the object of the war.

So, too, in naval warfare, the trial and error method of actual operations has dictated the selection of the hostile armed forces as the primary objective. Naval forces found that they could neither safeguard friendly lines of communication nor deny hostile lines of communication so long as the hostile armed forces were in existence and free to act. Surface forces found it necessary to defeat other surface forces before they could secure their ultimate aims: the will to resist.

Gentlemen, all this talk of "will to resist and will to fight" seems highly theoretical. What has all that to do with the soldier—who is a man of action? As a matter of fact, it has everything to do with the soldier. Even the clash of arms on the battlefield is essentially a clash of wills. The opposing commanders are fighting a battle of wills, and the first intimation of defeat or victory finds expression in the minds of those commanders. Even the private soldier is fighting essentially a battle of wills. The machine gun on the hill stops the progress of a hundred men. Why? None of them has been hurt. They are not shackled or chained to the ground. The terrain in front of them is not impassable. They are *physically free.* If they all go forward, some of them will get through—enough of them to dispatch the crew of the machine gun. And yet they do not go forward. They are stopped. They are held securely in their tracks, and the chains are none the less binding because they are invisible. They are held back by a powerful force—and that force is not physical but mental. Their will to progress has been temporarily overcome.

In the ultimate decision on the battlefield, defeat is acknowledged not by the maimed and wounded. Their influence as individuals has passed. It is acknowledged by the remainder—who have not been physically hurt. In modern war—all conflict is essentially a conflict of wills.

If this is true of the armed force on the battlefield, how much more powerful is its influence on untrained civilians. The soldier is carefully prepared for his conflict. He is inured to hardship. He is disciplined. Every effort is made to bolster his morale—his will to fight. War hysteria is carefully nurtured and fostered. He is led to believe that his sacrifice will guarantee the security of those at home. If he dies, it will be on the field of honor, and a grateful government will care for his family. Truly, the soldier is sustained by his pride—the visible manifestation of his morale.

Contrast this with the civilian. There is scant glory in watching his family grow emaciated from lack of food—in seeing them suffer for lack of heat—in walking the streets endlessly in search of work in a community whose factories are idle for lack of raw materials or power. None of the props which bolster the soldier's morale are present to support the will of the civilian.

And yet the loss of that morale in the civilian is just as conclusive as the defeat of the soldier on the battlefield. Small wonder that nations have maintained armies to shield that sensitive national heart—the national will to resist.

Gentlemen, the introduction of the airplane has wrought a profound change in the means of waging war. The ultimate end remains the same. It is and always has been the breakdown of the hostile will. But the means of accomplishing that end have been profoundly influenced. No barrier can be interposed to shield the civil populace against the airplane. A new instrument for the furtherance of national policy has been wrought. Old rules and methods cannot be applied blindly to it. It is impossible to hold position in the air. It is impossible to fortify a line in the air. The air is simply too large, too expansive, to be treated as a terrain feature. The air force can penetrate at will and proceed directly to its true objective. It is not necessary to fight an exhausting war in an effort to penetrate a barrier. We have found, not a useful new weapon to be used as an adjunct to the old; not a new projectile to be included in the family of supporting fire weapons; but a *new means of waging war*. Surface warfare *requires* an initial conflict between armed forces; the air force can strike at once at its ultimate objective: the national will to resist.

The air force is at liberty to proceed directly to the accomplishment of the ultimate aim in war: overthrow of the enemy will to resist through destruction of those vital elements upon which modern social life is dependent.

The ability to operate initially and directly in furtherance of the ultimate aim in war differentiates the air force from surface forces, and is the basic factor in the introduction of a new means of waging war. The air force itself differs from land and sea forces in three particulars. In the first place, air forces traverse any kind of terrain with equal facility. Rivers, mountains, lakes, or seas form simply interesting changes in the panorama below. Such terrain features are, of course, primary factors in the conduct of surface warfare. In the second place, the air force is not confined to its own element in the prosecution of war. Land forces fight only on the land. Sea forces fight only on the sea. The land force is unable to fight sea forces or to control sea areas. Naval forces are unable to fight armies or to control land areas. Surface forces are not only forced to accept the enemy's armed forces as their strategical objectives: they must exert their efforts against particular kinds of armed forces. The air force, on the other hand, operates with equal facility against objects either on the land or on the sea. It is not forced to accept the enemy's armed forces as the principal objective, and, if air forces *do* operate against enemy air forces by choice, they can attack *any* form of the enemy's forces.

The third particular in which air forces differ from surface forces lies in the fact that air forces do not have to *occupy* in order to exert pressure. Surface forces are constrained to accept occupation as the *only* means of expressing their force, of achieving their ultimate aim. Air forces are not restricted by this limitation. It is not necessary to physically occupy an area in order to render it useless to the enemy.

Those three characteristic advantages are peculiar to the air force, and together contribute to the one *great advantage* of air warfare over older means of waging war: the ability of air forces to proceed at once to the ultimate aim. It is the part of wisdom to exploit that great advantage to the utmost. To be sure, the air force may be used to influence the conduct of surface war by surface forces. But the air force contributes its maximum effort toward successful conclusion of the war by exploiting to the utmost the advantages peculiar to air warfare.

Gentlemen, we stated earlier in this discussion that armed forces are instruments for the furtherance of national policy. Armed forces do not shape national policy. The national need for prosperity, security, or racial unity dictates the composition and size of the armed forces.

Even in cases where the political head which shapes the national policy and the military head which fashions and employs the armed forces are fused into one personality—as Napoleon and Frederick the Great in the last century, or Mussolini, Hitler, or Stalin in the present—even under those circumstances it is the policy essential to peace which dictates the policy for war. The national *need* for prosperous peace is the basis for the design of the instruments of war. The soldier is servant to the statesman.

The national *need* may be divided into three general categories. The national *need* may require *physical acquisition; political acquiescence;* or *physical and political defense.*

That is to say, national need may require the taking and keeping of territory belonging to another nation, and the maintenance of political domain over the people of that territory; as an example, the need for raw materials and the internal pressure of over population caused Japan to take Manchuria and Italy to conquer Ethiopia; it may force Germany to expand in the valley to the Danube and conquer a new colonial empire.

Again, national need may require political acquiescence. It may require one nation's accepting the policies of another—without any thought of permanent acquisition of territory or political right. For example, American participation in the last war. Our participation in that war was prompted largely by the desire to force the Central Powers to accept American policies. We had no need for additional territory—but we *might* need extension of existing policies, particularly commercial policies.

The third national *need* is for maintenance of existing policies, and continuance of political control over existing dominion. In other words, the need to fight to maintain what we already have in the face of a covetous and aggressive foe. This seems to be the principal need for the maintenance of American armed forces at this stage of our national development.

To accomplish these needs three instruments are available today: ground forces, sea forces, and air forces. Each has peculiar powers. They should be apportioned relative to each other in light of their peculiar powers toward accomplishing the national need.

For the acquisition of territory, Armies furnish the final evidence of success. Air forces may assist by direct action against the national will, or by destroying or emasculating the means to fight. Navies *may* assist by safeguarding the transportation of men and supplies against enemy sea forces. But the final consolidation of victory and establishment of civil government may require physical occupation, and devolves upon the Army.

In the other two types of national need, occupation is incidental to the breaking of the hostile will. It is temporary. It may or may not be necessary, depending upon whether sufficient pressure may be brought to bear in other ways. Certainly where frontiers are contiguous, armies will play an important role in defending against other armies.

Where frontiers are *not* contiguous, armies must be transported to the theatre of war in order to exert influence. If air forces are within range to permit of strategic offense, they may be the decisive factor. Whether or not both air forces are in position to take the strategic offense, air forces will always be in position to affect the transportation of surface forces to exposed frontiers.

In the latter two categories of national need: the need to force acquiescence and the need to maintain defense, the principal reliance may devolve upon the air force.

To summarize our conclusions, gentlemen: War between major powers is an effort to settle by violence a conflict of national policies. The *aim* in war is to force an unwilling government to accept peace on terms which favor our policies. Since the actions of that hostile government are based upon the will of the people, no victory can be complete until that will can be molded to our purpose. The ultimate aim of *all* armed forces is to break down the will to resist.

For the accomplishment of that *ultimate aim*, three types of armed forces are available: land forces, sea forces, and air forces. Each has powers and limitations peculiar to itself. Those powers and limitations govern the application of the means to the end. Land forces and sea forces express their powers through *occupation*. The *ultimate objective* of land forces is the occupation and control of vital enemy territory. The *ultimate objective* of sea forces is the occupation and control of vital sea areas and sea lanes of communication. The *ultimate objective* of air forces is the destruction of the vital elements within the enemy nation.

The limitations of surface forces impose the acceptance of intermediate objectives. Since strategy is the *application* of a means to an end, let us call these intermediate objectives *strategic objectives,* since the strategy of surface warfare is necessarily concerned with achievement of those preliminary ends.

The *primary strategic objective* of land forces is defeat of the hostile land forces.

The *primary strategic objective* of sea forces is to defeat or contain the hostile sea forces.

Air forces are not constrained by the same limitations. Their primary strategic objective is normally identical with their ultimate objective. Air forces proceed directly to destruction of vital elements within the enemy nation in order to break the enemy's will to resist.

Gentlemen, we are confronted with the challenge that air forces have never proved themselves in the crucial laboratory of war.

Let us accept that indictment—not reluctantly, but quite honestly. To be sure, the air force has never won its spurs. That is not surprising. It came of age after the last tournament had ended and before the next has begun.

It is true that the air force has never demonstrated its power to change the age-old methods of waging war by orthodox means. However, it is equally true that neither armies nor navies have proved their powers in the face of a real air force.

Let us not be zealots. Let us not plunge thoughtlessly from the old and known to the new and untried. Let us not claim that the airplane has outmoded all other machines of war. Rather, let us be content with an evident truth: The air force *has* introduced a *new* and *different means of waging war.*

Alexander Pope left us a couplet. The first line has been quoted to us many times: "Be not the first by whom the new is tried." Very well. But let us not forget the rest of that sage advice. "Be not the first by whom the new is tried, *Nor yet the last to lay the old aside.*"[21]

Comments on "The Aim in War"

The first two-thirds of the lecture expand on the lectures of George, Fairchild, and Wilson by discussing specific economic and racial causes for war and explaining why wars have been fought as they have in the past with the only forces available, that is, armies and navies. In the final third of the lecture, Hansell introduces the reason why the ACTS believes air power to be inherently offensive in nature by suggesting that the vastness of the air domain makes defense against aerial attack infeasible. In 1936 he cannot anticipate how radar will invalidate this key assumption for strategic bombing in just a few short years.

Nonetheless, Hansell does articulate here a theory of victory that is distinct from that requiring battlefield defeat and occupation. While armies may be the appropriate means to achieve the brute force objective of annexing territory, air forces may be capable of victory by coercing political concessions or deterring enemy aggression. These important concepts of coercion and deterrence remain relevant today as air power strategists continue to evaluate how air power may be utilized independently to achieve limited aims.

Together, these three lectures and Harold George's lecture in chapter 1 provided the intellectual foundation for the Air Corps Tactical School's theory of strategic bombing. Fairchild emphasized how peacetime thinking led to the creation of air power. Wilson stressed how the advent of the airplane had changed the means to wage war. Finally, Hansell highlighted how air forces, unlike surface forces, could directly target the ultimate objective of war: the will of the enemy population to resist by destroying the vital elements of a nation.

Given this theoretical framework, the next two chapters reveal the ACTS offensive operational construct whereby victory can be achieved by large formations of self-defending bombers that penetrate enemy defenses and conduct high-altitude daylight precision bombing.

3

The Bomber Always Gets Through

Whereas the previous two chapters considered theoretical questions regarding air power in future wars, this chapter addresses the tactical challenge for bomber formations to penetrate enemy air defenses. Two lectures, "Driving Home the Bombardment Attack" and "Tactical Offense and Tactical Defense," offer distinct arguments as to how and why self-defending bomber forces without the protection of dedicated air cover could successfully conduct deep strikes in enemy territory without suffering debilitating losses.

DRIVING HOME THE BOMBARDMENT ATTACK

Lieutenant Kenneth Walker argues that a well-armed and well-motivated offensive bomber force can penetrate and strike its target in the face of enemy defenders and without the aid of air cover.[1] This article, a summary of the ACTS 1931 bombardment text, discusses the attributes of different flying formations that allow a massed bomber group to defend against enemy interceptors and anti-aircraft fire.[2] Walker acknowledges that, while a bomber formation will likely take casualties, training and the determined will of the commander will propel the bombers to their destination.

Driving Home the Bombardment Attack
Kenneth Walker
October 1930

When a bombardment unit clears its airdromes with a mission of destroying a vital objective deep within a hostile territory, it will be opposed vigorously by the enemy's defense forces; the hostile pursuit aviation and antiaircraft artillery.[3] The unit will be confronted with a task no more difficult than that

Kenneth Walker, circa 1930.

which confronts the infantry when it jumps off on a well-planned and coordinated attack. As the infantry receives the support of other ground arms, so does the bombardment unit receive the support, either special or general, from the other classes of aviation—observation, attack and pursuit, necessary to drive home the bombardment attack. In examining the tactics which a bombardment unit will employ to insure its arrival over and the attack of the objective, it will be found that this class of aviation operates at high altitudes and low; by day and by night; in formation and by series of single airplanes.

A brief review of bombardment organization and equipment may assist somewhat in the understanding of the present accepted bombardment tactics. Bombardment aviation is organized into squadrons, groups and wings. Two or more combat squadrons with a service squadron and other auxiliary units compose a group; two or more groups with necessary auxiliary units compose a wing. Each squadron is equipped with thirteen airplanes, of which a maximum of ten are expected to be always in commission. The group, with four combat squadrons, for example, is expected to put a maximum of forty airplanes in the air.

The number of squadrons which will operate against a particular target will depend upon the type of objective and the hostile opposition expected. It is impossible to determine, in the abstract, the strength which must be employed against any particular type of target. It will, to use the overworked phrase, "depend upon the situation." However, bombardment tactics are developed with a view to the proper employment of whatever number of airplanes must be used to accomplish a mission, rather than being based upon specific types of objectives which bombardment aviation will attack. The tactics developed are adapted primarily to the squadron and group organization. When more than one group is employed against a particular objective, the wing tactics consist of one group guiding upon the other, maintaining such intervals as are necessary for coordinated action and mutual support.

Bombardment aviation employs two types of airplanes; the heavy bombardment airplane capable of carrying a 2,400-pound bomb load; the light bombardment airplane capable of a 1,100-pound bomb load. Ordinarily the units equipped with the heavy bombers will operate at night. Those equipped with the light bombers will operate in daytime. Each, however, are suitable for and may operate both day and night. While at the present time the heavy bombardment airplanes only are in service use, light bombardment airplanes are under construction. The heavy bomber must be capable of high speed of at least 125 miles per hour and must have a radius of at least 300 miles; the light bomber a high speed of 160 miles per hour and a radius of action of 200 miles. Each bomber is twin-engined—one engine placed outboard on either side of the fuselage. In the nose of the fuselage extending forward of the leading edge of the wings, are placed the pilot, the bomber and the front gunner. In the rear portion of the fuselage are located the rear gunners. An alternate arrangement of rear gunners is to place them in the engine nacelles. By such an arrangement, excellent vision for the pilot and the bomber are afforded. Flexibly mounted machine guns cover all areas open to the approach of hostile attacking aircraft. The airplane is equipped with radio telephone with which communications between airplanes may be maintained and formations controlled in flight. With the rapid advance of aeronautical development the above conception of the proper types of bombardment airplanes will be changed from time to time, when increased performance will make possible greater load and cruising range.

For purposes of discussion, bombardment tactics will be reviewed under the following headings:

Day operations at high altitudes.
Day operations at low altitudes.
Night operations.
Special support by other classes of aviation.

Day Operations at High Altitude

In conducting day operations at high altitudes—meaning altitudes above ten thousand feet—a bombardment unit will normally perform a mission with its airplanes in formation. The formation lends itself to the delivery of a mass attack, to defensive machine gun fire superior to that which may be brought against it, and affords a measure of security against antiaircraft fire. That a mass attack is delivered from a formation is, of course, obvious. It is necessary, however, to investigate the types of formations adopted to understand their defensive powers when opposed to hostile pursuit and antiaircraft artillery.

Although it is not desired to consider in detail all points concerning the bombardment formation, it is believed that the discussion which follows will indicate sufficiently the features upon which the foregoing statements are predicated. First, the formation must be simple, compact and capable of ready control by the formation leader. It must be capable of maneuver and so flexible that distances and intervals between individual airplanes may be readily opened and closed. Its arrangement must be such that all angles of approach by hostile aviation are well covered by defensive machine gun fire. In this connection, emphasis is placed upon a formation arrangement whereby the maximum fire may be concentrated against that angle most favorable to attack by single seater pursuit. The formation must be so flown that a simultaneous attack by a superior number of hostile pursuit is difficult.

To meet these requirements, the normal formation consists of a number of three or five airplane elements. Within elements the airplanes are echeloned rearward from the leading airplane to the right and left and slightly upward in altitude, forming a V. Each element flies to the rear of the preceding element. The elements are echeloned *downward* from front to rear. With such arrangement all areas enclosing the formation are well covered by machine gun fire. By the "staggered down" feature embodied in echeloning elements downward from front to rear, all rear gunners are provided with unblanketed fire to the upper rear hemisphere, which is the angle of approach most favorable to the attack of hostile single seater pursuit. To appreciate the fact that approach from the upper rear hemisphere is most favorable to hostile pursuit, consider the difficulties of the frontal or flank attack, or the attack from the lower rear hemisphere of a formation. In the frontal attack, the speed of approach of the pursuit is the sum of the speeds of the pursuit airplanes and the bombardment formation. This great speed limits the time in which the attacking pursuit is in position to deliver accurate aimed fire to but a few seconds. In the flank attack the target is moving at right angles to the line of fire of the pursuit—aimed fire is again difficult. In the attack from the lower rear hemisphere, the pursuit airplane pulls up from a dive beneath the formation. The speed of the pursuit airplanes is materially reduced in the upward climb and the airplane "hangs" beneath the formation within range of the bombardment machine guns a longer period of time than is available for the pursuit airplanes to deliver aimed fire. In an attack from the upper rear hemisphere, the speed of approach is the difference between the speed of the formation and the attacking airplane; the front guns of the latter may be aimed from the beginning of the dive to completing the attack; the speed built up in the dive insures rapid withdrawal upon completion of the attack; the formation is moving generally in line of fire of the attacking pursuit airplane. Pursuit will attack from all angles, however, and as above noted, all angles of approach are well covered by machine gun fire, but with the maximum gun fire available to the upper rear hemisphere.

As the upper rear hemisphere is most favorable to pursuit attack, the formation is as narrow laterally as is consistent with concentration of defensive machine gun fire, to make difficult the simultaneous attack by large numbers of pursuit airplanes from this angle. Thus, a bombardment group formation of four squadrons can be easily flown within an area 500 feet wide and 1,000 feet long. It will be most difficult for an equal or superior number of pursuit airplanes to launch a coordinated, concentrated attack against a group formation of this character. Even though 40 pursuit airplanes could deliver a simultaneous attack against such a formation, it would be bringing but 80 machine guns into action against either 160 or 240 guns mounted on 40 bombardment airplanes. As the rate of fire of the flexibly mounted machine gun is nearly twice that of the machine gun mounted to fire through the propeller, it is apparent that the bombardment formation should have the best of the argument, by sheer force of fire power.

Pursuit will attack by long range fire, as well as by close range fire. A

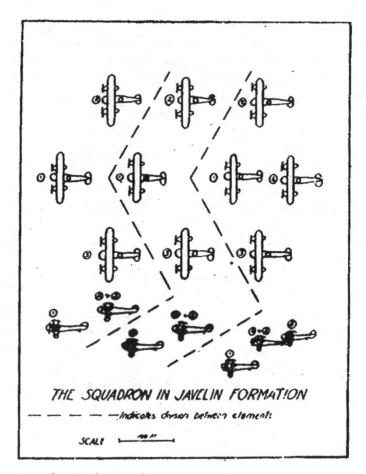

A ten-ship Bomber Squadron in a Javelin formation (overhead and side view).

group formation as compact as that above discussed is undesirable, in that machine gun fire, delivered in the plane of the formation, may miss the airplane at which it is aimed, but hit another airplane. When such fire is anticipated the bombardment formation may increase interval and distance between airplanes to from one to two hundred feet in from about one to three seconds. A hostile pursuit force may attack with a number of airplanes flying to the rear of the formation and delivering long range fire, while other airplanes deliver close range fire, approaching the formation from several angles. The open formation will be used against such an attack. Accuracy in fire will be an important factor in the relative number of pursuit and bombardment airplanes hit. The fact remains that the bombardment formation is still delivering a superior volume of fire against the attackers.

Pursuit may employ a time-fused fragmentation bomb which may be dropped on a bombardment formation from above. Two-seater pursuit is being developed. A hostile force equipped with airplanes of this type, may form on the flanks and in front of the bombardment formation, and concentrate against it the fire of the flexibly mounted rear guns. While the bombardment formation is the recipient of either or both of the above types of attack, other pursuit may attack the formation from the rear with their fixed guns. For the defense against this type of attack, support by friendly pursuit may be required.

A formation designed for defense against hostile aircraft is not entirely suitable for the avoidance of antiaircraft gun fire. A compact defensive formation is less maneuverable and it provides a larger target against which all antiaircraft batteries within range may be concentrated. That formation most suitable for operations over areas defended by antiaircraft artillery consists of one in which the airplanes are flown with considerable intervals and distances, i.e., where the airplanes are dispersed rather than concentrated. One type of dispersed group formation, known as the "dispersed column," is cited to illustrate. In this formation each squadron will have ten airplanes, the normal number. The airplanes are flown in two elements of five airplanes each, one behind the other. When opening up to a dispersed column, the leading squadron maintains the lower altitude. The second and third squadrons take positions on the flanks and above to the rear of the leading squadron, each maintaining a distance of approximately one thousand five hundred feet from the leading squadron. The fourth squadron flies to the rear of the leading squadron at a distance of approximately 3,000 feet and about 2,000 feet above the leading squadron. Within squadrons, the second elements are echeloned upward in altitude to the rear of the first or leading element. The individual airplanes are flown from four hundred to six hundred feet apart in their respective elements. Within such a formation, the airplanes are constantly changing speed, altitude and direction in maintaining the assigned distances. When antiaircraft fire is anticipated or experienced, each airplane, guiding upon the one in front of it within its respective element, engages in decided maneuvers. Endeavor is made to change altitude, speed or direction, or a combination of these, within the time of flight of the antiaircraft shell to the altitude at which the airplanes are flying. With these distances between airplanes, one antiaircraft shell can injure but one airplane. Should all batteries within range concentrate on one squadron, the other squadrons are not in danger. If the batteries do not concentrate their fire, the probability of hits is reduced. When attacking a compact defensive formation, all batteries may concentrate their fire against *the formation*, with the probability that slight errors in fire, directed against a particular airplane, will hit another airplane in the formation, and that a shell which hits or detonates near one airplane may seriously damage another airplane.

In a group formation such as described above, forty individual and separate targets are presented to the antiaircraft artillery. By plotting an antiaircraft gun defense, the area in which effective fire may be delivered is of course determined. The time during which the formation will be within range of the batteries may be calculated. A formation flying at a speed of from two to three miles a minute will be within effective antiaircraft ranges but for a short space of time. These tactics present a problem to the defending antiaircraft artillery far greater than that presented when the bombardment formation approaches an objective in a compact formation.

In bombing from such a formation, each airplane is held to a straight course for those seconds (not to exceed twenty) required to perform the timing operation and release the bombs. Upon release of bombs the airplanes again assume a maneuvering course until the defended area is passed through.

When a bombardment unit takes off to perform a mission, it will normally open to a dispersed formation. The compact defensive formation is required only for defense against hostile aircraft. The bombardment pilots are subjected to less strain in flying the dispersed formation than in the defensive formation. The route selected for the mission will avoid, as nearly as possible, the known or suspected areas in which hostile pursuit is certain to be operating, and where antiaircraft artillery is sure to be emplaced. If, however, there is no alternative, the formation will proceed to the objective through the hostile combat zone. As it is possible that the hostile pursuit will attempt to intercept the bombardment formation as it crosses the hostile front lines, the air force commander will arrange that, at the time and place where the bombardment unit crosses the lines, friendly pursuit will be present in force. By such action, the bombardment formation may be enabled to maintain the dispersed formation while flying over the combat zone. Should hostile pursuit be present, the friendly pursuit should be able to effectively prevent

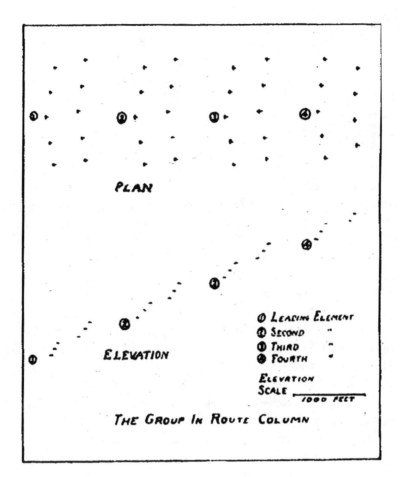

A thirty-nine-ship Bomber Group in a Route Column formation (overhead and side view).

the former from attacking the bombardment formation. A combat zone of 20 to 30 miles in depth may be crossed in from 10 to 15 minutes. Unless the enemy pursuit succeeds in engaging the bombardment formation without being prevented by the friendly pursuit the bombardment should be able to traverse the combat zone in dispersed formation and thus limit the effect of antiaircraft opposition. If, however, the hostile pursuit is present in force and is not prevented from attacking the bombardment formation, the latter will assume a defensive formation. It is unlikely that hostile antiaircraft will fire when its own pursuit is present and engaged in attacking the bombardment formation. A coordinated attack by pursuit and antiaircraft would be difficult of accomplishment without considerable danger to the pursuit. The antiaircraft doctrine, which in effect is that when friendly pursuit is present in force, the antiaircraft artillery withholds its fire, is logical and will doubtless be applied.

Upon passing through the combat zone, the bombardment route will avoid the antiaircraft batteries grouped around vital points in the system of rail communications, important supply establishments, etc. Should hostile pursuit be absent or prevented by friendly pursuit from attacking the bombardment formation when the bombardment formation crosses the hostile front, it is expected that hostile pursuit units in the air and on the alert at airdromes will be notified of the presence of the bombardment formation. These pursuit units will endeavor to intercept the bombardment formation as quickly as possible. The time required for interception will be that necessary for transmission of information to the pursuit units; the time required to issue orders, clear the airdrome and climb to the altitude at which the bombardment formation is flying, if the pursuit unit be on the ground; the difference in speeds between the bombardment formation and friendly pursuit; and the accuracy with which the hostile pursuit units carry out the interception. Should interception by hostile pursuit be effected before the objective is reached, the bombardment unit will assume a defensive formation and engage in a running fight until the objective is reached. If the pursuit force fails in preventing the bombardment formation from reaching its objective, but continues attacking the latter when it arrives within range of the antiaircraft artillery, the bombardment unit will assume a dispersed formation, release its bombs therefrom, and be prepared to close up to a defensive formation upon clearing the range of the antiaircraft batteries. The time required for a bombardment formation to open or close, to assume one formation or another, is measured in seconds, rather than minutes. It may appear wise in theory for the antiaircraft to refrain from firing when a dispersed formation is assumed, on the proposition that pursuit will then dive in to engage the bombardment airplanes, or for pursuit to refrain from attacking a defensive formation on the assumption that antiaircraft artillery will then open fire. However, it is submitted that the extreme nicety or coordination of such tactics will cause delays which, measured in time, will be such to allow the bombardment formation to proceed a great deal of time without being subjected either to the fire of pursuit or antiaircraft. If the action of pursuit and antiaircraft can be so coordinated and perfected that antiaircraft can fire during the intervals between successive pursuit attacks, without danger to the attacking pursuit, it may be then habitual for friendly pursuit to support a bombardment formation. Friendly pursuit should be able to break up any coordinated attack by the hostile pursuit, thus permitting the bombardment unit to maintain an open or dispersed formation without sacrifice of the scheme of defensive machine gun fire. . . . [The sections "Day Operations at Low Altitude," "Night Operations," and "Special Support by Other Classes of Aviation" have been omitted.]

Conclusion

An attempt has been made to outline, in a most general way, those methods by which a bombardment attack may be driven home. There are many alternative methods which are believed practicable but which follow to an extent those above discussed. How efficacious the tactics may prove to be can only be discovered in war. It is believed, however, that proper application of these tactics will lead to successful bombardment operations.

By no means may it be assumed that bombardment units, applying these or any other tactics, can avoid casualties. A certain loss in men and material is a price which must be paid for success. This is true for any military force. The results obtained will be the determining factor as to whether or not the price paid is too high.

It must be remembered by those responsible for the defense against bombardment operations, that a bombardment unit will not be stopped by the presence of a strong defense or a mere show of force. It is generally conceded, by those who are competent to judge, that an air attack well launched is most difficult to stop. The bursting of antiaircraft shells or the presence of a hostile pursuit force will not prevent a determined bombardment commander from accomplishing his mission. To stop a bombardment attack, the bombardment airplanes must be shot out of the sky.

In the final analysis, the most efficacious method of stopping a bombardment attack is to destroy the bombardment airplanes before they take to the air. As a bombardment unit will be upon its airdrome *at least* sixteen out of every twenty-four hours, the *best defense* would appear to be *an offensive* against the bombardment airdrome.

Comments on "Driving Home the Bombardment Attack"

Compared to the other ACTS lectures presented in this book, this piece is the most dated because it was published in 1930, five years before the arrival of the four-engine B-17, a platform that doubled the air speed and combat range Kenneth Walker envisioned in a two-engine heavy bomber. Walker's central argument is that bombers in tight formation would have the advantage in both the number and rate of fire of its machine guns. Fighters, even in large formations, would have to break up their formations to prosecute attacks, thus conceding the advantage to the bomber.

Over time, Walker's conviction regarding the offensive nature of air power became a key assumption in ACTS strategic bombing theory and underpinned the belief that a tightly flown bomber formation could defend itself against any enemy defense. Of all the ACTS faculty, Walker was the most zealous, a prophesier who based his views purely on faith rather than on evidence from previous wars. In his conclusion, he attempted to insulate his claims from criticism by claiming his views were accepted by those "competent to judge." Though he does not name who those are, we are left with the impression that such competence resides with him and his bomber brethren alone.

Interestingly, Walker was largely proven right in World War II as the bomber formation almost always did get through. Even on the most infamous raids in which US bombers suffered horrendous losses, such as at Schweinfurt and Regensburg, the bombers made it to the target. Walker, however, addressed only the effectiveness of a single bomber raid while ignoring the question as to whether there would be sufficient aircraft available for subsequent strikes. The success of air operations was borne out by the cumulative effect of attacks over time, not on the outcome of a single raid. The central question that Walker did not address, and which is considered in the next lecture, was whether a bomber formation could attack its target with an acceptable loss rate.

Tactical Offense and Tactical Defense

Major Frederick Hopkins, in his 1939 lecture on "Tactical Offense and Tactical Defense," addresses this critical issue of bomber attrition.[4] From the evidence of the British Independent Force's daylight air raids in World War I, Hopkins concludes that only when the ratio of pursuit to bombers exceeds 1.5 to 1 does the attrition rate become too high for sustained offensive operations. Hopkins argues that such a ratio would prove rare as the offense would inherently have the advantage over the defense as the offense chose the time and place of attack. The offensive could mass its force at the point of attack, while defenders, unaware of exactly when and where the point of attack would be, had to disperse its forces and wait. The concentration of offensive force would thus prevent the pursuit to bomber ratio from becoming too high and attrition rates too great.

Tactical Offense and Tactical Defense
Major Frederick Hopkins
March 31, 1939

Gentlemen: In his conference last hour Major Thomas referred to the "penetration of artificial resistance."[5] It is our purpose to explore this subject at this time. From the lectures presented thus far in the Air Force Course, it must be evident that we should have a clear conception of the characteristics of Air Forces. No decision affecting the organization and employment of our air forces made in time of peace is more important than that to be made from the conclusions which are to be drawn from these conferences this morning.

The capacity of a nation to supply its armed air forces the necessary service aircraft in time of war or before the advent of war is not unlimited. It was pointed out last hour that someone in peacetime must decide the proportion of defensive pursuit aircraft we will need on M-Day. Someone in peacetime must decide the number of striking force airplanes which will be needed *unless* we find it possible for a nation in adopting the strategic defense to depend 100% upon defensive action to protect those vital establishments which we believe an enemy would feel justified in attacking.

Insofar as is possible we will endeavor to *compare* the air tactical offensive with the air tactical defensive; to balance the air offensive against the defense that can be provided by pursuit aviation and against the resistance which can be interposed to counter offensive action. We will confine ourselves to air tactics eliminating all strategic considerations.

We will attempt to be perfectly reasonable. We have been investigating this morning the characteristics of an air force with a view to exploiting its particular powers. In considering the air offensive and the antiaircraft defense we will, however, try to retain a perfectly unbiased point of view. Let us emphasize that we are investigating a question that is being debated by the foremost military writers among the leading air powers of the world.

If we consider just the air combat phase, there are those who relegate the defensive pursuit airplane to second place, contending that this type of airplane has little chance of success in attacking a striking force airplane safeguarded by all around fire; that striking force airplanes which can concentrate their mutually supporting fire when flying in close formation are practically invulnerable to pursuit attack. There are those who contend that the speed of the larger airplane has approached that of the pursuit airplane, making interception difficult and evasion easy, and, that while it is true that a striking force airplane presents a larger target, yet it can be exposed to a larger number of hits before being put out of action.

On the other hand, we have those who claim that the modern pursuit airplane armed with cannon is fatal to bombardment airplanes flying in close formation; that the efficiency of the weapons carried by the striking force airplane will be considerably limited at altitude and by the high angular velocity in the brief moment available when firing at a modern diving pursuit airplane; that the pursuit airplane approaches and passes at the tremendous speed in angular velocity of 187 degree per second which is entirely too fast for the flexible guns of the defensive formation to fire with any degree of accuracy.

We are sure that neither of these views is correct.

We are not sure of how best to establish any basis of comparison.

On the inside cover of the Air Force Text of the Army War College there is a quotation attributed to Lieutenant General von der Goltz. This quotation may be found in the preface to the Fifth Edition of the works of this illustrious gentleman, a work which he wrote in 1883, entitled *The Nation in Arms*. This quotation reads: "It has been my endeavor to trace the developments in the range of military operations *in conjunction with the older ones* and *to determine their influence on martial events*. Progress along such a course may be less dashing and rapid than a surrender to the flight of fancy and a visionary forecast of the coming events, but the prospect of safely reaching one's goal will certainly be much brighter."[6]

This quotation is interesting because it is a practical line of approach to this discussion. If we follow this sage advice of studying the old, we should evaluate the lessons of the World War insofar as they apply to our subject—modifying such evaluation by considering the trends in such operations in wars since the World War.

It is, however, rather strange to add that this Prussian General whose service extended from the Franco-Prussian War to the eve of the World War surrendered to a flight of fancy by advocating that fortresses be made mobile by the use of prefabricated parts of iron and steel which "form a whole and which are capable of transference from place to place."

And perhaps as a visionary forecaster of the coming of air power and mechanization he prophesied in 1883: "The day will come when the present aspect of war will dissolve, when forms, customs and opinions will again be altered. Looking forward into the future, we seem to feel the coming of a time when the armed millions of the present will have played their part. A new Alexander will arise who with a small body of well equipped and skilled warriors will drive the impotent hordes before him."[7]

It is apparent that our concern will be with broad generalities rather than with specific combats or specific types of equipment. This is at best a difficult and controversial subject and if we permit ourselves to become involved in argument over combat tactics we will soon mire down and lose sight of our broader purpose.

In studying the old wars, however, we are disconcerted by a statement that often appears in military writings to the effect that a new war is never like the last war. General Kitchener is reported to have said of the trench warfare of the World War "that whatever it was, it wasn't war!"

Here we found stabilization developed to its highest degree. It was a war of attrition with anchored flanks from the straits to the frontier of Switzerland. It gave birth to air power. For air, however, it was chiefly a ground force cooperation type of war—it was a war in which the airplane simply intensified the killing power of the ground arms by supplying better maps, better information and better regulation of artillery fire. It was a war in which only the smallest percentage of airplanes could be made available for use in an air offensive. It is our contention, therefore, that although we may draw certain lessons from the World War we must be very careful in any application of these lessons to the next.

We have before used the British 41st Wing as the subject of a conference. It is to this force and to its expansion into the British Independent Force, RAF, that we are compelled to turn at this time, for it is the only independent offensive operation of which we have records that provide any kind of an analysis of the independent use of air power during the World War.

You may remember that towards the latter part of the war, a British bombing force was used against objectives in the Saar and Rhine basins. It was used to wage an air offensive against the German industrial structure. All of the elements of the antiaircraft defensive were present on the German side. The British bombing squadrons were trying to penetrate this artificial resistance. The German antiaircraft artillery and barrage balloons defended the objectives that were selected for British air attack. *We have here all of the basic elements upon which to base a rather fair discussion of the tactical offense and the tactical defense.*

Let us review then the general situation in order that we may give these air operations the tactical consideration that is quite necessary. The British Independent Force consisted initially of the 41st Wing of three squadrons—the 55th Day Squadron and the 16th and 100th Night Squadrons. Later, upon expansion into the VIII Brigade and, subsequently, into the Independent Force, three additional squadrons were added—the 99th, 104th and 110th Day Squadrons. The wing was based initially at Ochey but was moved later to Tantonville. *We find the Night Bombardment squadrons encountered little opposition from pursuit and antiaircraft artillery.* The German records indicate that only a single night bomber ran afoul of the German balloon apron defense. The greatest difficulties were encountered in the night navigation to the objective in the wretched weather of that theater of operations.

It is, therefore, in the employment of the Day Squadrons that we may expect to find the greatest result in our research.

All of the Day Squadrons were initially equipped with DH-9 airplanes. These airplanes were rather slow. They were approximately twenty miles an hour slower than the German Albatross and Fokker fighters that were encountered. The radius of action of these bombers was such that weather conditions, particularly the high westerly winds prevalent in that theater, had to be considered in all operations involving objectives in the Rhine Valley. Squadron personnel took three to four weeks of final service training in formation flying at their airdromes on the front before being sent over the lines. The air attacks against Germany were conducted by units of six to twelve airplanes in formation. Each airplane had one defensive gunner, firing in the upper rear hemisphere. There was limited gun fire to the front from the single fixed gun.

The number of German pursuit airplanes that opposed these British formations varied. At times, however, there was a German superiority of five to one in this area. From August, 1918, onwards 16 Home Defense flights of 15 aircraft each assisted by five Pursuit flights of 18 aircraft each had assembled along routes most likely to be followed by British aircraft. A total of over 300 first class fighters—Albatross and Fokkers—were defending an area which would include the Montgomery-Birmingham-Atlanta Sector. It is to be observed that all of the penetrations into Germany passed through a relatively narrow arc. The German pursuit forces were concentrated behind the lines along that narrow arc.

What, then, are the questions—the answers of which we are seeking?

We would this morning—if we could—call to our assistance the first leader of the air offensive—Viscount Hugh Trenchard, retired Marshal of the British Air Force, who commanded this Independent Force. We would like in our most polite manner to ask him these questions:

1. What percentage of striking force airplanes evaded German Pursuit?
2. What percentage of bombers which were exposed to the tactical defense were shot down in air combat?
3. What percentage of bombers were lost when the ratio of pursuit airplanes to bombers in air combat was one to one—two to one—three to one—four to one?
4. On what percentage of missions were alternate objectives bombed? And why?

It is possible to reconstruct from reports and squadron histories data to answer many of these questions. One squadron leader with tremendous pride in his outfit took time after the war to put into writing the complete story of the exploits of his squadron. Major L. A. Pattinson, commanding the 99th Day Squadron, has left us an invaluable record as source material. Would that there had been more such squadron leaders.

The problem of tactical pursuit employment must for the sake of analysis in this discussion be divided into two phases. First, the bringing of the two forces into air combat; and, second, the actual air combat. The first represents the problems that are involved in pursuit interception. The second involves the numerical superiority of pursuit which is needed to defeat the striking force bombers either by shooting them all down, or by making them aban-

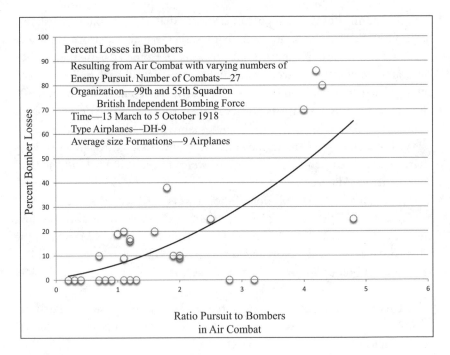

don their mission or by inflicting such casualties that the bombers fail to return.

It is our intention to discuss this latter phase—the actual air combat phase—first.

From the records of the Independent Force, RAF, as presented in the official history of the war, a tabulation of factors has been made—the objective, the number of bombardment airplanes employed, the number of German airplanes encountered, the ratio of German pursuit airplanes to British bombers in each combat, the number of bombers lost, the percentage of bombers lost and the results of the combat insofar as the mission was concerned. It has been impossible to arrive at any figure on the number of German pursuit airplanes that were shot down except from British reports. By analyzing this table we may gain some idea of the influence of pursuit numerical superiority upon the percentage of bomber losses.

The curve that has been drawn from this tabulated data will not satisfy those of you with engineering tendencies. As an ordinate on the vertical scale we have plotted the percentage of bomber losses. On the abscissa on the horizontal scale we have plotted the ratio of pursuit to bombers in combat. Keep in mind that the points plotted indicate actual air combats. A rather doubtful curve has been drawn through the points thus plotted but it may indicate *certain tendencies* that may be of value to us in our analysis. It most certainly indicates that bomber losses increased as the pursuit numerical superiority increased. It indicates that when German pursuit superiority was from four or five to one, that almost 100% losses could be expected by the invading bomber formations. It indicates that the DH-9 with its single flexible machine gun covering a quarter of the firing hemisphere could expect from zero to 20% losses when matched with an equal number of pursuit.

We should try to arrive at some general conclusion as to the pursuit numerical superiority in air combat that may make air force operations relatively unprofitable. Let me repeat that statement because much of our further discussion rests upon this point.

Let us try to set up some kind of assumptions. It was the experience of the World War and is still the basis of our War Procurement Plans that an average airplane replacement rate of 30% a month will be needed to sustain the efforts of our Army Air Corps. We sincerely hope this figure is too high for the operations of a striking force but due to the lack of better values we shall accept it. The logistics tables of this school and adopted by the GHQ Air Force are built upon an expected average of two-thirds of a mission a day, or twenty missions per month, a figure which we may believe high but which we will also accept without discussion.

What do these figures mean? They mean that we must keep the average losses of all kinds in twenty missions in a month to less than 30% or else exceed our replacement rate. These figures mean that whenever we exceed 1½ average losses per mission *of the total force* we are exceeding our replacement rate. They mean that if on the twenty missions per month if we exceed 30% losses *of the total force*, future air operations must on the whole be cut down and we are jeopardizing the sustained effect of air force employment. *It is rather obvious that if on any series of missions the entire striking force suffers more than 10% losses on each mission that air operations will soon be very seriously curtailed.* German air operations ceased against London when losses built up to 14%.

We will use this figure of 10% as a basis for further discussion. If we glance at our chart, we believe that you will concur that the British could expect such 10% losses when a striking force formation was outnumbered about 1.5 to 1. Such a figure taken from this chart would mean that when a

British formation of nine DH's was attacked by fourteen German Albatrosses the British usually lost one airplane.

We have previously mentioned a tabulation that was made for data for the preparation of this chart. There are several interesting additional figures that may be quoted.

A total of 191 day missions were conducted on a true air offensive into Germany. A total of 1,641 airplanes invaded Germany to bomb German objectives. This does not include the missions that were assigned in counter-air force operations against German airdromes. It does not include those forced to return because of engine trouble. An additional 228 airplanes could have bombed if engine trouble had not forced a return to the airdrome. This is a high percentage—17.5% of those that actually were exposed, and indicates the engine unreliability in 1918. We would not expect certainly such a percentage in any air operations that we would conduct in 1939. It includes only those which were exposed. Now, there were ninety-one bombers actually shot down in air combat. This is 5.5% of the number of bombers that started for an objective and were not forced to return for any reason. There were, of course, many other British losses that we are not interested in—crashes, forced landings in friendly and in enemy territory due to navigation or engine trouble—only one airplane is reported shot down by antiaircraft artillery fire. We note from British reports that these defensive formations that lost 91 bombers in action, shot down 43 German pursuit airplanes and drove 86 others down out of control—a total of 129. The greatest losses in bombers occurred when the German pursuit was built to a ratio of about five to one during the summer months of 1918. It is rare that we find these British pilots abandoning their missions except in bad weather. There is sufficient evidence that poor tactical judgment often resulted in casualties so high as to render complete squadrons impotent. There is not a bit of evidence that the British did not dare to return as soon as given new equipment and a new levy of inexperienced pilots.

At the time of the Repeal of the 18th Amendment, there was an eminent statistician who calculated that if the annual national output of beer was poured into a ditch four feet deep and twenty feet wide dug from the Great Lakes to the Gulf of Mexico that it most certainly would ruin the beer. We are about in the same position as this statistician. We may take this figure of numerical superiority of 1.5 to 1 simply as a general figure upon which to base further discussion. We know such a numerical superiority led to about 10% bombardment losses during the World War in the situation which we have

Table 3.1
DAY MISSIONS
British Independent Force RAF
(No Counter-Air-Force Operations)
Squadrons No. 55, 99, 100, 104, 110
OCT. 17, 1917–NOV. 10, 1918

ACTUAL AIR COMBAT	
No. of missions	191
No. of bomber airplanes exposed to air fighting in formation	1,641
Bombers shot down	91
RATIO = Bombers exposed	91 / 1,641 = 5.5%
German Pursuit Losses	
Driven down out of control	86
Shot down	43 *(From British Reports)*
German Pursuit Losses	129
ACTUAL INTERCEPTIONS	
No. of German Pursuit Interceptions	67
Percentage of Interceptions	67/191 = 35%
ANALYSIS	
Average no. of bombers in formation	1,641 / 191 = 8.6
Av. Bomber Losses on Interceptions	91 / 67 = 1.4
Av. Loss per Mission Intercepted	1.4 / 8.6 = 16%

been discussing. However, to attempt to reduce the factors of air combat to figures by use of a slide rule is rather absurd. It ignores the courage and valor of the Guynemers, Boelckes and Richthofens and their equally courageous bombing opponents.

We divided pursuit employment into two parts for the sake of analysis; first, the air combat phase, and second, the bringing of the two forces into combat. Let us examine this second phase. We should investigate the factors involved in Pursuit Interception.

There is some very interesting data indicating the number of interceptions made by German Pursuit of British bombers in 1918. In 191 day missions, German Pursuit forced air combat on British bombing formations sixty-seven times, or about 35% of the total. This is roughly a third and indicates one interception in three. Bear in mind the tremendous pursuit in this area but also a probably inefficient warning service. On an average of two out of three missions the British suffered no losses at all. In sixty-seven missions the British lost 91 airplanes or an average of 1.4 airplanes per mission. Since the formations averaged about 8.6 airplanes per mission this would indicate a loss of about 16% per mission when pursuit aviation was able to force combat on the penetrating formation. Certainly these are very serious losses.

Is there any reasoning then that we can use to arrive at any conclusions as to the percentage of the missions that will completely evade pursuit and suffer no losses at all? Any answer to this question will depend on the proverbial situation.

We have no way of evaluating this factor. We can set general maximum and minimum limits only. In the case of strong defensive effectiveness, such as the attack of a well defended air base with all local defense agencies with an efficient interception net, it is not unreasonable to expect a very high percentage of the striking force missions intercepted. Once air combat has been forced upon such a force we can expect maximum losses. On the other hand, in the attack of objectives where an aircraft interception net cannot be made available, such as in the defense of a coastal city or a naval vessel, we certainly expect a very small percentage of the striking force penetrations to be intercepted on the advance to the objective and must figure on the effects of interception after the bombing has been completed.

After a little serious study of this defense problem it is difficult to remain unbiased.

There are certain factors which we believe will hinder the tactical defensive from obtaining any such combat numerical superiority as 1.5 to 1, when we set up a certain set of conditions. Air offensive operations on the whole cannot be made unprofitable when we consider a tactical defense on any kind of a broad scale. We believe that there are three factors of sufficient importance that alone will definitely prove this contention; *first,* the influence of weather on both the air tactical offensive and on the air tactical defensive; *second,* the influence of the relative speed between bomber and fighter; and, *third,* the influence upon the defense of time, position and the number of objectives to be defended.

Let us consider the first—weather. Weather is a natural resistance—and outside the scope of this lecture, but let us examine it in the light of the operations of the British Independent Force. It exercised a decidedly different influence on the tactical air offensive in France than it would today. If we examine the records of the Independent Force, RAF, we find that 32% or one mission in three failed to reach the primary objective and bombed alternative objectives—alternative objectives that either were designated in orders or selected by the tactical judgment of the air combat commander. The principal reason for this failure to bomb primary objectives was due to the state of technical development of the airplane proper. It is not unfair to say that the World War ended before the airplane had developed its full power as an instrument of war.

The airplanes of these Day Squadrons were used to bomb at about the extreme radius of action under most favorable conditions. Given weather conditions of clouds, fog, rain, sleet and particularly high winds, the squadrons were grounded. Weather forecasting had not reached its present state of development and often after taking off, conditions arose that required the abandonment of deep penetrations into the Rhine Valley. The high winds of thirty miles per hour reduced the speed of the DH-9 about 30%—that of a present day bomber about 10%.

We believe that you will concur that no such conditions exist today. Such progress in blind flying has been made and is being made that probably if any of us were called upon to penetrate an active antiaircraft defense in daytime we would not want a clear and unlimited variety of weather. And if interceptions depended upon us as group commanders of pursuit aviation, undoubtedly we would pray for the cloudless skies that occurred day after day in sunny Spain. Weather has become a factor which permits evasion of the pursuit defense and will be a powerful factor in reducing striking force interceptions.

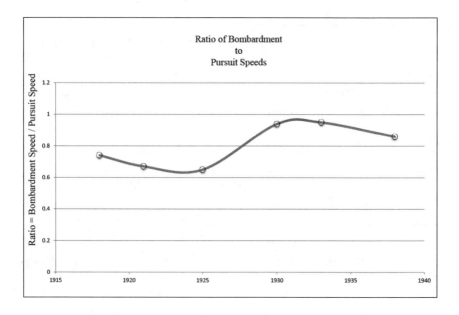

Our second factor is the speed superiority of the pursuit airplanes with respect to bombing aircraft. This is a most important factor. Very high speeds tend to affect seriously the maneuverability of the pursuit airplane. *However, the higher the speed of the bombardment airplane, the more favorable will be its use in the tactical offensive.* The higher the speed, the greater the aircraft interception net required for antiaircraft defense and, therefore, the fewer the vital establishments that can be provided a reasonable defense. The speed of these airplanes has almost tripled since 1918. *The effect of these high speeds is to reduce the time of air combat between any two points on the earth's surface.* These speeds require the fastest kind of estimation on the part of the air combat commander involved. Whereas in actual air combat, speed difference is of vital interest, we are interested particularly in the ratio of bombardment speed to pursuit speed when we consider the problems of interception.

Summing up, the ratio of bombardment to pursuit speed curve has had its downs and ups in the last twenty years. Any increase of the ratio of bomber speed to pursuit speed increases the difficulties of pursuit not only in interception but also in air combat, tending to reduce the effectiveness of contact. As a general rule newly designed and higher powered engines usually are placed in pursuit airplanes before being installed in the larger aircraft. It is the consensus of opinion of good airplane designers that given an engine, a four-engined bombardment airplane can be designed with a high speed very close to that of any single engine pursuit interceptor, to speeds well within 10%. It is not too much to say that future development at the Materials Division visualize a bombing airplane with a range of 4,000 miles, powered by four 2,000 H.P. engines capable of a high speed at 20,000 feet of between 380 and 400 miles per hour. It is recognized, therefore, that the single engine pursuit airplane must give way to a twin-engined interceptor type, say the XF-38, in order to regain the speed superiority necessary. These high bomber speeds have brought forth in Europe a new interest in the time fuse bomb for bombing bombardment formations and has aroused a discussion of the possibilities of ramming by pursuit.

There is the third and last factor that we must consider—the influence of time, position and the number of objectives to be defended. *Herein lies the weakness of the tactical defensive. This is the most important factor not only as regards the ability of pursuit aviation to intercept but also the ability of pursuit aviation to build a defense on any sort of a broad scale which insures a numerical superiority that would make air force operations unprofitable.*

Let us assume that we wish to defend a city with pursuit airplanes.

Let us assume that in conducting that defense we wish to engage all enemy forces in combat with a numerical superiority of 1.5 to 1, a figure which we will assume may result in 10% striking force losses. We accept this as merely a general figure.

The method by which we employ our pursuit airplanes will vary with the position of the city to be defended. Let us assume that the entire friendly territory within the frontier is organized into an aircraft interception net. As you will recall from your Pursuit Course, there are three general methods of pursuit operations—by patrol, by air alert, and by ground alert. Furthermore, there is a definite limit within which each of those methods can be employed. Objectives to be defended that are located on or near the frontier with no interception net extending toward the enemy require the use of the patrol method. If we move backward from that frontier we reach a distance at which it becomes feasible to use the air alert method, and still further back we reach a zone within which the ground alert method can be employed. The distances behind the frontier at which these methods can be used will vary with the performance of the friendly pursuit airplanes and the enemy attacking airplanes. The altitude at which the enemy attacking formations penetrate will also influence that distance.

If we assume that the city which we wish to defend is located back in the ground alert zone and *further assume the enemy has 100 bombers* which are within range of the city, we find that the enemy bombers have the tremendous advantage of initiative. They can select the time of day or night for attack. Since it is hardly reasonable to expect pursuit pilots to stay on the alert continuously for twenty-four hours without relief, day in and day out, let us recognize the fact that each pursuit airplane will require at least three combat crews. Would you, as pursuit commanders, ask for less?

Let us not lose sight of the fact that vast numbers of searchlights would be required for the defense of a large area at night—providing one of the most severe limitations to the employment of pursuit aviation. Let us even imagine that we have these thousands of searchlights.

Since within the ground alert zone we can bring all the available airplanes to bear against the enemy we find that we will need 150 pursuit airplanes and 450 combat crews to provide defense for this city against 100 enemy bombers. This is undoubtedly a minimum. The British defenses of London contemplate far greater numbers of pursuit aviation. It is possible, however, they fear far greater numbers in the attacking forces. So our ratio of 150 against 100 modern bombers is certainly not unduly optimistic.

If our city is in the air alert zone it will require more airplanes in order to maintain that 1.5 to 1 superiority. Let us remember that in the air alert method we keep our airplanes in the air. *Even the very thought of this method is fatiguing.* Since the airplanes must be maintained and the pilots must get some rest, we may assume that it will take three airplanes and six crews to keep one airplane in the air *continuously.* If that is the case we will find it necessary to provide *450 airplanes* and *900 crews* to bring about that same combat superiority of 1.5 to 1. If our city is up in this patrol area the problem becomes so difficult that we become convinced that some other means besides pursuit alone must be used to defend our objective.

Now, if the city represents the only element that must be defended within the enemy nation, such as London is assumed to be in the British Isles, then these figures are not necessarily exorbitant—particularly those for the ground alert. It seems quite feasible to establish a pursuit defense of a single locality that might prove highly efficient against an enemy striking force of about this size. Of course, if the enemy striking force increases in size, the problem becomes more staggering. Grant that the maintenance of 1,500 pursuit airplanes for defense against 1,000 bombers in the ground alert zone is quite feasible, reasonable, and economical, it is when we have more than one place to defend that we run into trouble. Because of the high speed of striking forces there is a limited distance to the front and side to which ground alert and air alert pursuit forces can make interception.

The same 100 bombing airplanes, that we initially assumed, might presumably attack any of, say, five cities. Hence under the best circumstances in which our vital elements are far enough behind our frontiers to permit us to use the ground alert method, and assuming that it is *possible* to use great numbers of pursuit planes in combat which it is not, especially at night, it would be necessary to provide 750 pursuit airplanes and 1,500 crews in order to gain the necessary 1.5 to 1 superiority in combat against the original 100 enemy bombers to protect these five cities. If those enemy bombers were 1,000 in number it would require ten times as many to meet those same conditions. *This type of reasoning is, of course "reduction ad absurdum." But the general inference cannot be escaped.* It may be feasible and economical to establish a defense of a particular locality that will effectively limit repeated air attacks. However, if the things to be defended are not concentrated, but rather widely separated, then it is not feasible or economical to provide that kind of a defense. *Since this latter seems to be the usual case, we can but conclude that to provide pursuit defenses on a broad scale that are capable of guaranteeing a defense which will make enemy air offensive operations unprofitable is not feasible.*

If we examine the wars since the World War, are there lessons which may modify these views? Certainly there is little to learn from the Italian-Ethiopian conflict insofar as air fighting is concerned.

In China, it is the consensus of opinion of military observers that the air operations of both the Japanese and the Chinese are far below the standards of the occidental powers. Compact formations have not been used by the Japanese bombing formations although it is thought that the theory of mutually supporting fires is taught at the Mannatsui Bombing School. Japanese night navigation has been excellent. The reports of the results of normal air combat between Japanese bombardment and Chinese pursuit, while the latter existed, vary considerably. Japanese bombers with blind angles about the tail—those with large twin rudders seem to have suffered severely. A report of the Chinese aircraft interception net established in the early days of the war was received from a former instructor of the school here as follows: "The net was organized shortly after my arrival in China, May 31, and naturally there was not time to train personnel properly or to test the net. In the first days of the war, the following message was received from one of the stations: 'There is a loud noise passing overhead going in your direction.'"[8]

Most of the operations in China seem to be on this sort of standard.

As for Spain, we have demanded some first hand information of the war from Captain Griffiss in a coming conference which we are looking forward to with a great deal of interest.[9] During the last summer of the war, aircraft of superior performance were used by both the Insurgent and Government forces. The superior pursuit armament of the Russian-built airplanes seems to be outstanding. The following report may be of interest to us in our discussion at this time:

> Though the actions of the air forces in Spain have been on a very small scale in comparison with what they would be in a war between large nations of Europe, they have nevertheless offered an interesting field for investigation and observation.
>
> First of all is the importance assumed by the factor of speed. It is *speed* that enables the airplanes to carry out their missions before they are overtaken by the pursuit airplanes of the enemy, to escape the fires of these fighting airplanes and that is to say, reach objectives that are more and more distant. The more the units have been equipped with slow equipment, the greater the losses they have sustained and the greater the difficulties they have found in carrying out their missions. They have been an easy prey for the pursuit planes and the modern antiaircraft guns. The use of new planes; the Soviet Katiuskas, the Junkers, the Heinkels, the Dornier 177 and the Savoias 79, whose speed is approximately 400 Km/hr (about 250 m.p.h.) almost equal to that of the pursuit airplanes has made it possible to recommence the long distance bombardment missions.

Let us definitely state our conclusions:

We believe that the trend in technical development has favored the offense at the expense of the defense.

The *initiative of the offensive* gives to the attacker not only the *choice of time* and *place of attack* but the amount of force to be used.

We know that under no conditions can the tactical defensive secure a favorable ultimate decision.

It seems very improbable at this time that we can *guarantee 100% on a broad scale* an effective denial of enemy air attack by means of either pursuit defenses or other active agencies.

We must not conclude that it is not feasible to establish adequate pursuit defenses of a single small locality or area. We must not conclude that pursuit defenses are not desirable or that pursuit airplanes will not shoot down and destroy bombardment airplanes or that pursuit defenses are not essential. Not one of these latter conclusions is justified.

But we are forced now—in time of peace—in this discussion of the characteristics of air forces, to make a decision. Our decision can only be that of the defensive limited aim—to interpose that pursuit force—and only that force, the threat of which, requires for the enemy—the adoption of a maximum in security measures.

Later in the course we will be required to defend a large vital area against a striking enemy. Your conclusion will be that the best defense that can be provided is a *bitter* air offensive against those attacking forces which threaten us.

For those of you whom we have not convinced here this morning we simply ask you to hold your judgment in abeyance until you are called upon to solve that problem.

Are there any questions or comments?

Comments on "Tactical Offense and Tactical Defense"

Hopkins's inductive, evidence-based approach to offensive bomber operations was both logical and conducive to several useful insights. His finding that a 10% attrition rate was the maximum allowable for sustained offensive air operations was validated in World War II, Korea, and Vietnam. Hopkins also highlighted a key problem for defenders: the ability to obtain a favorable ratio of fighters to bombers when the bombers had the initiative.

Unfortunately, neither Hopkins nor his fellow ACTS faculty appear to have been aware of the implications the recent advent of radar would have on air defenses. Hopkins's lecture was, in fact, delivered just as the British were completing the installation of the early-warning radars to their Chain Home air defense system, an addition that proved critical to the RAF Fighter Command's victory in the Battle of Britain.

Together, these two lectures provide the justification for the ACTS belief that the offense would dominate the defense in the air domain. Air Corps efforts should therefore be focused on designing and building large numbers of bombers at the expense of pursuit aircraft. The next chapter considers the operational requirements in terms of bombers, bomb load, and bomber accuracy for conducting high-altitude daylight precision bombing.

4

High-Altitude Daylight Precision Bombardment

The ACTS faculty did not anticipate the impact radar would have on the offense-defense balance for aerial combat. Instead they worked on the operational requirements for a heavy bomber, made manifest in the B-17, to effectively conduct high-altitude daylight precision bombardment. In combat, bombers were compelled to fly at high altitudes due to the lethality of enemy anti-aircraft fire. This adjustment, however, introduced the significant challenge of bombing accuracy. Only a highly precise bombsight, such as the Norden Mark I, could deliver bombs from near-stratospheric altitudes with the requisite precision. Estimating the accuracy of the average bombardier, the challenge for air planners became that of determining the number of bombs and bombers to assign in order to destroy a target with a reasonable degree of certainty.

PRACTICAL BOMBING PROBABILITIES

The lecture by Captain Laurence Kuter, "Practical Bombing Probabilities," given at the conclusion of the Bombardment course, examines the lessons learned from the bombing probability problems assigned during class.[1] The key factors that determine the likelihood of hitting the target are the number of aircraft flown/bombs dropped, the altitude of weapons delivery, and the accuracy of the bombsight. Kuter argues that improvement in bombsight accuracy is where the greatest gains could be achieved.[2]

Practical Bombing Probabilities
Captain Laurence Kuter
February 13, 1939

Laurence Kuter, 1935.

This discussion concludes the portion of this course devoted to that phase of bombing accuracy entitled "Practical Bombing Probabilities."[3] In the preceding three periods and in our two Practical Bombing Probabilities Problems, we have not only found the potential power of Bombardment Aviation tremendously restricted by bombing accuracy (better called bombing inaccuracy in this case), but we have also uncovered some valuable suggestions as to means of remedying this very severe limitation. In this period we will briefly examine the results of those problems. Those answers themselves teach us certain important facts; those answers also show us various methods of correcting these severe limitations. We will examine those methods and determine which is the most practicable method of reducing this severe limitation. We find that method to be increasing the accuracy of bombing. We will discuss various possibilities of increasing bombing accuracy, outline the steps that have been taken, present the results that have been obtained and emphasize the action that must be taken in the next year.

We will first examine the results obtained in these problems and discuss the tremendous reduction in the potential power of Bombardment Aviation imposed by bombing inaccuracy.

High-Altitude Daylight Precision Bombardment 119

In the first requirement of Practical Bombing Probabilities Problem No. 2, we were concerned with the attack of five canal locks. Each of these objectives could be destroyed by one 300-lb. bomb. Five 300-lb. bombs could have accomplished this mission. Five 300-lb. bombs are ⅓ of the maximum bomb load of one B-17 or B-18 airplane. Five 300-lb. bombs would be an easy load for a Keystone, or for an old wooden Martin, and yet the solution to this requirement indicates the use of 88 B-10 type airplanes or ten bombardment squadrons to accomplish the mission which is within the potential possibility of one Keystone airplane. There is an almost overwhelming limitation. This limitation is based upon the mean probable error obtained in Table 4.1. These targets are small. Their dimensions are about 60 × 150 feet. These targets are smaller than almost any naval objective Bombardment might be required to attack. However, these targets are larger than many bridge piers which might very possibly be bombardment objectives. If they were placed in the center of a bombing range, these areas—60 × 150 ft.—would be big areas, and yet the mean probable errors in Table 4.1 demand that we use ten bombardment squadrons. Ten bombardment squadrons are more than the GHQ Air Force could put in the air at one time today. Only ten squadrons would not appear unreasonably large to Mussolini, to Hitler, or to Stalin. This particular objective may be vital to the ability of the United States to prepare for, wage, or sustain war. In time of national stress this force, even to us, might not appear unreasonably large. It is, today, a big force. Such a mission is beyond the capability of all of our Bombardment Groups today.

If some changes could be made so that a smaller proportion of our bombardment force could accomplish such a mission, we would have, in effect, greatly increased the strength of our whole bombardment force. As to the size force required against this objective, whether or not it is reasonably or unreasonably big, we must do everything in our power to take the necessary action so that we are not forced to employ so great a force on such a target.

In these two Practical Bombing Probabilities Problems we have investigated three methods of reducing the size force required (or of increasing the effective strength of Bombardment forces available). Each of these three methods is effective. Only one of these three methods will be found practicable.

In Practical Bombing Probabilities Problem No. 1 we investigated one of these methods: Reduction in altitude. In that problem we found that by reducing altitudes or bombing at more favorable altitudes, fewer airplanes were required (or the effective strength of available airplanes was increased). We found that we must drop 76 bombs to be practically certain of hitting that carrier from 18,000 feet altitude. By reducing our altitude 33%, or bombing at 12,000 feet (a more favorable altitude), only 27 bombs were required. In that case we reduced our altitude by 33%; we reduced the force required by 66%. Some such ratio with these errors will be found to be correct in general. A reduction of 25% altitude will increase the potential effectiveness of this bombardment force by much more than 25%, 45%, or 50%, for example. This is, therefore, an effective method of reducing the size force required. It is manifestly not a practicable method. We came down in altitude 25%; our bombing accuracy was increased by 45%. What was the increase in the effectiveness of the modern antiaircraft artillery guns with that decrease in altitude? The answer to that question will be known only when antiaircraft artillery guns have fired at current bombardment airplanes in a war between progressive, well-equipped, major nations. In the meantime, considering security only, if we reduce our altitudes, such an action will be wholly to our disadvantage if antiaircraft artillery effectiveness increases in greater proportion than bombing accuracy increases. We will never know where these two curves intersect until we have returned from missions and counted our losses. Those of us who do not come back will provide the data on which the surviving bombardiers can draw these curves. Until that date we know that reductions in altitude which seek most efficient bombing conditions cannot be the *reliable* means of correcting bombing inaccuracy.

In the third requirement of Practical Bombing Probabilities Problem No. 2, we have investigated a second method of reducing the size force required. There we have investigated reductions in the percent chance of success demanded of our bombardment units. We found that a 90% chance of success required 440 bombs. This School accepts the 90% chance of success as a practical certainty. However, there may be bombardment commanders who will not accept odds of nine to one as practical certainty. Had we demanded odds of 95 to 100, we would have increased our chance of success by 5%. We would have increased the number of bombs required to 600, or 36%. With a 95% chance of success, there would still remain five chances, in a hundred, of failure. It is conceivable that some individuals might not tolerate even so small a chance of failure. Had we demanded a 98% chance, we would have been required to drop 800 bombs. Here an 8% increase in chance of success demands an increase of 82% of the force employed. We encounter here the vivid conclusion that we cannot be *sure* of success without paying heavily for overwhelming odds. Had we been satisfied with an 80% chance of success, our force required could be cut 27%—down to 320 bombs. If we could have accepted a 50% chance of success, we might have reduced our force 69% and dropped 138 bombs. Here it is evident that a reduction in percent chance of success required will permit a proportionately greater reduction in size force requirement. When we go down 10% in chance of success we reduce our force 27%. This is an effective method of reducing the strength requirement. Unfortunately, this is obviously an impractical method of avoiding our bombing inaccuracy restriction. We cannot recommend accepting worse odds as a method of increasing effectiveness. Those two methods—choosing favorable altitudes and not demanding high orders of success—are both effective, and unfortunately, are equally impracticable.

In the first requirement in Practical Bombing Probabilities Problem No. 2 we investigated the effect of a drastic reduction in mean probable error. We investigated the effect of a drastic increase in bombing accuracy. We found that a reduction of mean probable errors to $\frac{1}{3}$ their former amount justified a reduction in size force required from ten squadrons to $1\frac{2}{3}$ squadrons. Here our mean probable error was cut to $\frac{1}{3}$ and our force requirement was simultaneously cut to $\frac{1}{6}$. Had we cut our mean probable error to $\frac{1}{4}$, our force requirement would have been cut to $\frac{1}{14}$. This, then, is the most effective method of reducing the strength requirement or of increasing the effectiveness of our available bombardment force. This method is a material improvement in bombing accuracy and here, fortunately, is one method that is practicable. Let us investigate the possibility of materially reducing the errors listed in Table 4.1, investigate the steps that have been taken to do so, the results that have been obtained in those steps, and the steps still to be accomplished.

This severe limitation, which demands that we use eighty-eight airplanes to obtain five hits, which are within the potential power of one airplane, is directly responsible to the mean probable errors listed in Table 4.1. Our interest lies in a reduction of those errors.

Many of us feel that we may take advantage of our accumulated experience and arbitrarily reduce the errors listed in that table. All recent Air Corps tests, of which any records have been kept, have shown considerably smaller errors than those listed in Table 4.1.[4] In the bombing of the ships off the Virginia Capes in 1921 and 1923; in the bombing for the McNair Board

Table 4.1. Established Probable Errors

Altitude (ft.)	Circular (Cep)	Range (Rep)	Direction (Dep)	Mean Probable Error* (μep)
5,000	142	87	82	84
6,000	172	103	97	100
7,000	201	120	115	117
8,000	231	137	131	134
9,000	261	153	147	150
10,000	290	170	163	167
11,000	320	187	180	184
12,000	350	203	196	200
13,000	379	220	213	217
14,000	409	237	229	234
15,000	438	254	245	250
16,000	468	271	261	267

* This column for practical use in conjunction with employment of the Table of Probability Factors.

Source: Air Corps Tactical School, *Bombardment* (Maxwell AFB: Air Corps Tactical School, 1935), table 1, AFHRA, A2686: Pursuit & Bombardment Texts, 1925–1940, 1168.

in Hawaii in 1929; in the bombing of the Pee Dee River bridge in 1927; in the bombing in the Joint Army-Navy Exercises off Langley Field last Fall; and every other time when the bombardment component of the Army Air Corps has been forced into a conspicuous responsible bombing test, the errors have been considerably smaller than those listed in the Table 4.1 that was current at the time of those tests. Arbitrarily reducing those errors would be a neat but not effective method of talking ourselves out of the hole this table has placed us in. This table is the Air Corps' story and we are stuck with it.

This School, and eventually the Bombardment Section, is asked almost weekly how accurate the Mark I bomb sight is, how many bombs it takes to hit a battleship, and other very short and simple questions. The source of these very clear, very direct, and very embarrassing questions is frequently some imposing personage ranging from the Commandant of the Command

and General Staff School or the Army War College to the Chief of the Air Corps.

By this time, you will know that the answers to these questions are: (1) Nobody knows; (2) Depending upon the situation, the number of bombs required to hit anything varies from a very small number to a very large number. These obvious answers are the only ones that have not been given to these questions.

Until two or three years ago, such questions were answered by reference to this Table 4.1, which has appeared in all Bombardment texts prior to the one issued to this class.

These errors tabulated in this old Table 4.1 are a precise statement of the errors actually measured in seven annual bombing matches at Langley Field, Virginia. Old Table 4.1 is the only record of D-1 and D-4 accuracy. This bombing was done by picked and carefully selected teams. One team was selected from each bombardment unit in the United States and Panama to compete for the gold "Distinguished Bomber" medal. It is possible these teams were not always the best bombing teams in existence, but they were well above average. Considering this as an accurate statement of average bombing errors, personnel and training considerations would most certainly indicate that these errors were not "average" but much smaller than average.

These errors, from the consideration of training only, should have been multiplied by two, three, or by four to get an average statement. Considering altitudes, this bombing was accomplished at 5,000 and 8,000 feet, only. These two particular altitudes—5,000 feet and 8,000 feet—in the opinion of the antiaircraft artillery enthusiast, might well be called the "suicide altitudes." As far as altitude is concerned, again, it would appear that these errors should have been materially increased. Considering the methods of approach, the bombing in these matches was performed in series of six bombs. Each bombing team approached the target six times at the same altitude, from the same direction, and under the same conditions. These teams came down this same groove six times, dropping one bomb each time. The first bomb dropped was called the sighting shot. The errors of the sighting shots were not even recorded. This table was, therefore, built on the errors determined by five bombs, dropped under identical circumstances, one after the other, completely disregarding the errors of the first in the series of six. In tactical bombing we envisage the bombing teams approaching the target, perhaps under fire, making one quick sighting operation, dropping bombs, and rapidly leaving the zone of fire. This tactical bombing is most certainly a far cry from the conditions under which the bombing recorded in these tables was accomplished. Again, from this viewpoint of method of bombing and method of approach, there should have been a very great increase in these errors.

In discussing weather conditions, visibility, turbulence of air, and so forth, we know that all of this bombing was performed in the daytime. We know that the Annual Bombing and Gunnery Matches are conducted in conditions of at least good wind condition and very good visibility. On at least one occasion these bombing teams waited for several days for these conditions to arise.

Considering the airplanes and their speeds, from which this particular bombing was accomplished, and the airplanes and speeds that we now discuss in bombing tactics, we find the old airplane a large, sturdy ship, cruising over the bombing range at from 65 to 80 miles per hour. The bombardier had all the time in the world to accomplish the sighting operation. We now talk freely of speeds well over 100 miles an hour in excess of that bombing speed. Once more, considering speed in airplanes, we can more justly increase these errors than arbitrarily reduce them.

All bombing done in this old Table 4.1 was accomplished with the D-1 and D-4 bomb sights. These bomb sights were technically entitled "the pendulous timing sights." The bombing we are doing now and the bombing we count on in the near future, is accomplished by the Mk. I bomb sight, a gyroscopic synchronizing sight. Is the Mk. I bomb sight, in itself, sufficiently more accurate than the D-1 and D-4 bomb sights to outweigh all these considerations—training, altitude, approach, speed, airplane, and weather—and still permit us to emerge with the belief that we can hit anything at all? We believe that it is. Fortunately, we will not be forced to make arbitrary reductions in an old table. We are building new tables.

Let us here take up one of the questions most frequently asked. How much more accurate is the Mk. I bomb sight than the old D-1 or D-4 bomb sight? You have seen that the only basis on which to compare D-4 is a far cry from tactical bombing. We have *one* test which we believe may be accepted as conclusive in providing an answer to this question.

Lieutenants Kilpatrick and Sullivan, in the 2d Bombardment Group at Langley Field, were enthusiastic, intelligent, interested bombardiers. When the Mk. I bomb sight was first made available to them, they mounted that bomb sight in a Keystone airplane and proceeded to bomb under exactly the same conditions that these Annual Match bombing errors were determined. While these two were, at that time, above average, they had had very little practice with this bomb sight. They were without doubt below average in skill of average bombardiers today. Comparatively they were far below the average of the bombing teams in the Annual Matches. As a result of their bombing we may make the following statement: "Under practically identical conditions of bombing, circular probable errors actually determined with the D-4 bomb sight have been reduced by the Mk. I bomb sight, at 5,000 feet, from 142 feet to 21 feet, and at 8,000 feet, from 238 feet to 28 feet. On an average these errors have actually been reduced to ⅛ the previous errors. This average represents a reduction in Probable Error of 87.2 percent."

Another and more comprehensive statement may be made of the relationship of current bombing accuracy with former bombing accuracy: "Bombing errors at 5,000 feet, with the Mk. I bomb sight, at high speeds, in the B-10 airplane, are about the same as they were with the D-4 bomb sight in the Keystone airplane." At 8,000 feet our errors are actually less than one-half the former efforts; at 12,000 feet, at high speed, we are actually obtaining errors only ⅔ as great as those we estimated we would obtain with the D-4 bomb sight in the slow Keystone, if we could have bombed at that altitude. The past history of the Air Corps bombing is obviously a sorry one. The past history of Air Corps interest in its ability to bomb is still sorrier. This may be due to a certain extent to lack of incentive or lack of instruction from higher Army authorities who did have general data on exterior ballistics. The first interest shown by the United States Army, outside the Air Corps, is dated May 7, 1938. It consists of a student study from the Air War College. The title of the study is "The Recording" by Lt. Colonel H. R. Harmon, who was, of course, Air Corps.[5]

We will devote the remainder of this discussion to process of obtaining reliable bombing data and the progress of that project.

A new TR 440-40 is now in effect. The first bombing seasons under the provisions of that regulation have been completed. The results of this bombing have been received and are being studied and analyzed. That bombing avoids the old, unsatisfactory, "down the groove" method of approach to bombing as follows: The bombardier approaches the target from some cardinal point, passes, say south, drops one bomb, turns; approaches from the east, drops one bomb, turns; approaches from the north, drops one bomb, turns; approaches from the west and drops his fourth bomb. That completes his bombing at that altitude. There, instead of disregarding sighting shots, we have nothing but sighting shots. Those may be considered approaching the tactical situation, at least from the consideration that one sighting operation only is performed, one bomb only is dropped under an identical condition of altitude and direction of approach. These Training Regulations will also avoid the selection of only two comfortable altitudes within probable effective antiaircraft artillery altitudes in the following manner. The four directions of approach at one altitude that we have just discussed will be at 4,000 feet. The bomber will then climb 4,000 feet, repeat those four approaches at 8,000 feet, then repeat the approaches at 12,000 feet and at 16,000 feet and so forth to the service ceiling of the current type bombardment airplane.

The maximum altitude is not arbitrarily set at some figure that the next model airplane may not look upon as extreme, but the maximum altitude is listed simply as the service ceiling of the current bombardment type airplane. After this Training Regulation has been in effect for some time it is evident that we will have accurate data on bombing, under conditions both of altitude and direction of approach very similar to those that may be encountered in actual service. The most recent Training Directive from the GHQ Air Force expands the instructions contained in this Training Regulation. The GHQ Directive calls for considerable formation bombing, night bombing, bombing in bad weather, bombing in conditions of poor visibility and at the conclusion of long missions. It is apparent that we will have truly valuable data after a great number of bombs have been dropped. A most important point which has not been overlooked by this Training Regulation and GHQ Air Force Directive is that data on this bombing must be accurately prepared and must be received, compiled, studied and interpreted at one central point.

The situation which results in these statements in Training Regulations and in Training Directives has been developing over the past four or five years. During this time, for lack of a better "central point," the Air Corps Tactical School has been designated as the point to which all bombing records would be sent for analysis and study.

In these past two or three years some of you have been involved with the preparation of bombing records and have forwarded them to this School. Others of you may have been involved in the preparation of these records but have not bothered to forward them to this School. That neglect has not made much difference. Those scores have not been properly analyzed at the Air Corps Tactical School. They have been referred to the Bombardment Section where we have made a few simple incomplete analyses. We have made one effort toward analyzing the effect of speed on bombing accuracy.

This first, feeble investigation discloses a tremendous increase in range probable error as the air speed of bombing increases from 100 to 200 miles per hour. This Section made two or three other simple similar analyses, probably the most important of which is represented by Table 4.1 in the text.

These studies fell far short of properly studying bombing records. We could not, or, at any rate, we did not do that job properly.

However, we did not conceal these shortcomings from the Chief of the Air Corps. There were a series of official communications and numerous personal visits apprising the Chief's Office of the fact that very little of the potential value of these bombing records was being extracted by The Air Corps Tactical School. The last and culminating official communication on this subject was dated about two years ago. It is believed desirable to quote a few paragraphs from the important paper. It was labeled "Confidential" and addressed to the Chief of the Air Corps.

CONFIDENTIAL
B. PROJECT FOR ANALYSIS OF BOMBING RECORDS.
1. NECESSITY THEREFOR.
a. BOMBARDMENT TACTICS.

(1) *Effectiveness of force.*—That the effectiveness of a bombardment force varies with its bombing accuracy is immediately evident. The degree of this variation is far greater than initial consideration might indicate. For example, prior to the advent of the Mark I bomb sights, the best available data indicated that about 10 squadrons would have been required to obtain a practical certainty of destroying the locks in the Sault Ste. Marie Canal. Since the advent of the Mark I bomb sight, probable errors have been reduced to about one-third of their previous magnitude. With this increase in bombing accuracy, only 15 airplanes would be required to obtain the same results under similar circumstances. Here a reduction in probable error to one-third has increased the effectiveness of bombardment almost six times. In this particular example if the probable error is reduced to one-quarter the original error, the effectiveness of that bombardment unit would, as a consequence, be increased 17 times. With this example, it is immediately apparent that detailed informa-

tion on "M-day bombing accuracy" is of tremendous importance to the bombardment commander, and to the commander of any higher unit which is directing the employment of bombardment aviation.

(2) *Training of bombardiers.*—The variations in bombing probable errors now employed in TR 440-40 provide an easy means of rating the proficiency of bombardiers, and, in addition, have a practical tactical value not found in any arbitrary system of scoring.

(3) *Logistics.*—When the number and type of probable bombardment objectives are known, data on bombing accuracy are required to form the basis for estimating the quantity of bombs that should be provided bombardment units.

b. BOMBARDMENT EQUIPMENT AND TECHNIQUE.

(1) *Bombs.*—Bombing accuracy and the military characteristics of bombs are intimately associated. It is essential that bombing records be carefully studied, with special reference to dispersion and variations from the mean trajectory, to set up satisfactory military characteristics for bombs. Once these characteristics are known, the Ordnance Department is properly charged with the design, development, and supply of bombs with these characteristics.

(2) *Developments.*—Effective general coordination of the development of bomb sights, bomb racks, and the bombardment airplane with the employment of bombardment aviation, can be accomplished only by one central office which has immediate access to all detailed information relating to bombing errors.

2. PRESENT SITUATION.

a. ANALYSIS OF BOMBING RECORDS.—At present, bombing records are not analyzed. They are received at the Air Corps Tactical School as described in Section A of this Memorandum. The Air Corps Tactical School lacks both personnel of necessary scientific qualifications and personnel of necessary mathematical ability to conduct this analysis. As a result, if three hostile airplane carriers were to appear off our coast today, no Bombardment Group commander, no Wing commander, nor any higher military headquarters has any sound basis upon which to determine whether he should dispatch all his bombardment airplanes against one carrier, or whether each carrier might be attacked, or whether a small portion of the bombardment aviation available would be a sufficient force to accomplish the desired results.

b. BOMBARDMENT EQUIPMENT AND TECHNIQUE.—At Aberdeen Proving Ground an effort is being made to provide firing tables from the results of necessarily few bombs. The problem of preparing these tables is of considerably greater magnitude than preparing a similar table for a piece of artillery. Throughout the Army Air Corps tens of thousands of bombs are being dropped annually. These bombs could provide data from which firing tables of unquestioned accuracy could be produced. No study is being conducted to determine the source of the errors that compose the bombing probable error. No effort is being made to isolate these errors with a view to their reduction or elimination, and yet a reduction of these errors may operate to the same effect as doubling or tripling the number of the bombardment units now existing.

3. REMEDIAL ACTION.

a. In order to remedy the serious impairment to the efficient employment of bombardment aviation, and the logical development of its tactics, technique, and equipment, described in the preceding paragraph, immediate action should be taken to organize a section whose duty is the analysis of bombing records.

b. The duties of this section should be: (1) To determine the type and magnitude of current bombing errors. (2) To determine the causes of bombing errors and the means of their elimination or reduction.

c. The steps to be taken in the accomplishment of these two investigations are quite similar and can properly be undertaken by the same section. As this study is of a continuing nature, the results of which should be disseminated periodically, the personnel should be permanently employed.

d. The composition of this section should initially comprise a physicist with two clerical assistants. The qualifications required of the physicist are those of a high quality research analyst with especial ability in original physical research. The necessary qualifications of the two clerical assistants include the ability to strip and select data and prepare accurate curves therefrom.

e. The organization, direction, and supervision of this section should be a responsibility of the Air Corps Board. A considerable interest on the part of the Ordnance Department in the operation of this section of the Board is to be expected. The operation of this section of the Board bears directly on the Ordnance function of the design, development, and procurement of those bombs which are required by the Air Corps. Aside from this single aspect, all other findings of this section of the Board will pertain solely to the Air Corps. The Material Division will show considerable interest in the operation of this section of the Board when findings of the analysis of bombing records indicate shortcomings in bomb sights, bomb racks, and other items of bombardment equipment. Bombardment units will be particularly interested when findings point the way to material improvements in bombing training and technique. The Air Corps Tactical School, all bombardment units, and all other agencies concerned with the employment of bombardment aviation will be intimately affected by the findings of this section of the Board, through the practical tactical use of the bombing probable errors accurately determined.

4. RECOMMENDATIONS.

a. That the Air Corps Board be charged with the continuing responsibility of evaluating bombing records and preparing probability tables and graphs.

b. That immediate action be taken to provide the Air Corps Board with one physicist and two clerical assistants, whose duties will be evaluation of bombing results and analysis of bombing accuracy, for the dual purpose of providing information as to current bombing accuracy and analyzing the causes of error with the view to their elimination or reduction. (In this connection, it should be pointed out that until the services of a physicist are made available to the Air Corps Board, little or no evaluation of bombing results can be accomplished.)

5. The above study is the result of deliberations between several officers on the faculty of the Air Corps Tactical School and the members of the Air Corps Board, and to which considerable thought has been given over a long period of time with information investigations and consideration to the ideas of officers in tactical units, the

Materiel Division, and the Ordnance Department. The members of the Air Corps Board and of the faculty of the school are in unanimous accord with regard to the recommendations made herein.
CONFIDENTIAL

This paper did have the desired effect. Money was made available in the next budget and early this Fall Doctor Heinecke and his assistant arrived. The Research Section of the Air Corps Board has been set up and operating for about six months. Their office is in the basement of this building, just off the Reproduction Section's rooms and I am sure if you will find the office, Doctor Heinecke at any time will interrupt his work and explain what they are doing with courtesy, if not with pleasure.

Colonel Sorenson has taken a very active interest in the supervision of this Section.[6] Early in May he hopes to have in his hands the first specific and obviously worthwhile results from this section. At that time we plan on having Colonel Sorenson present to you the first findings of this Research Section.

Meanwhile we are concerned about impatience. We know that one should not allow the analysis of bombing results to drift for twenty years and then expect to make up for all neglected opportunities in the next few months. This Research Section costs the Air Corps several thousands of dollars per year. We know there are many projects that the Air Corps would like to carry on with such funds. Given time we believe this section may, to a large extent, affect a "marked reduction in mean probable error," not unlike that investigated in Practical Bombing Probabilities Problem No. 2. However, even if this section accomplishes no more than the equivalent of increasing the effective strength of the GHQ Air Force by one B-17 airplane, the funds saved by that accomplishment would be sufficient to sustain this Section for 25 years. We hope that you will hear more detailed information on this subject in May.

At this point we may conclude that we have not reduced the mean probable error as it probably could have been reduced in the past few years but we have not been asleep on this subject. With the continued receipt of usable, carefully prepared bombing records it may be possible in the immediate future to isolate some of the error contributing to the bombing probable error and to eliminate or reduce many of them.

Many of you have listened to or participated in recent Air Corps discussions of the "sighting shot." Particularly those of you who have been close to the Headquarters of the First Wing at March Field, California, are familiar with the good logic with which they support their contention that sighting shots must be employed. That logic is contained in this situation: Where a Bombardment unit leaves its base and proceeds a thousand miles to sea, flying through bad weather, climbing to very high altitudes to avoid some weather, when it eventually arrives in the vicinity of its objective, no bombardier in the entire organization has any means whatsoever of checking the effects of this long flight with great changes in temperature, possibly having run in and out of icing conditions several times, on his bomb shackles, bomb racks or any of his bombing machinery. Furthermore, he has no knowledge whatsoever of the ballistic wind. He may be bombing at 18,000 feet in a 100 mile per hour west wind. During the 36 seconds it takes his bomb to fall that bomb may be moving through winds totally different in direction and velocity. Under these conditions First Wing Headquarters contends that these bombardiers are in a very sad situation and the sighting shot must surely be employed.

The logic of that argument is sound and we believe that some people have been too hasty in not giving it proper consideration. For example, assume we have a mean probable error of 150 feet, we know that if we approach and bomb this point and if our bomb sight is operated perfectly, if racks, shackles and all bombing machinery function properly that there is a fifty-fifty chance that that bomb will strike 150 feet "over" the target. Assume that that bomb was our sighting shot and we correct our sights by that amount for the next bomb. We would, in effect, correct them by aiming at a point 150 feet short of the target. Our second bomb likewise has a fifty-fifty chance of falling 150 short over the point to which it was aimed. Hence this second and "corrected" bomb would strike 300 feet short and we would be much worse off than we were having no sighting shot whatsoever.

If two sighting shots had been dropped the probability is not remote that both of them might have been 150 feet over. Obviously, in order to use sighting shots effectively a considerable number must be dropped—that number has not yet been determined. It is understood that the antiaircraft artillery adjusts for ballistic wind effect after firing four sighting rounds, maybe four is our number, maybe we have to go to a much greater number.

Hence this sighting shot argument cannot be dismissed without careful consideration. We do feel, however, that our basic and fundamental conception of bombing accuracy must be based on a situation wherein Bombardment approaches an objective, probably under fire, takes one quick sight on the target, drops its bombs and leaves. We feel we must base our computations on that initial round and not on an adjustment of fire.

With respect to the ballistic wind we hope the time will come when that factor will demand our serious attention. Such is not yet the case, however, inasmuch as the contribution to the mean probable error made by variations in the ballistic wind is still the smallest single error that can be isolated.

We cannot intelligently conclude this discussion of bombing accuracy without reference to Spain and China.

CONFIDENTIAL
THE AIR CORPS TACTICAL SCHOOL
Maxwell Field, Alabama
MEMORANDUM: Bombing Accuracy in Spain and in China.

The Air Corps Tactical School has for some time believed that the "mis-use" and general ineffectiveness of bombardment aviation in Spain and in China has been due not so much to the poor judgment of the higher commanders as to the poor marksmanship of their bombardiers. In Spain, in China, as to a large extent in the latter phases of the World War, bombardment is used extensively to supplement the artillery barrage, to indiscriminately drop bombs over large areas of important cities and, in general, to operate at considerable variance with our concept of efficient employment.

We have believed that the Japanese and all contestants in Spain have found that their bombardment units have been unable to "cut sensitive points in rail lines of communication" for the same reason that the 1st Day Bombardment Group failed to cut the important Mezieres-Metz rail line; namely, bombing accuracy is so poor that an unreasonably large number of bombs must be dropped to obtain an effective hit.

We know that the Germans had practically uninterrupted service over the Mezieres-Metz tracks during the three months and three days when the 96th Day Bombardment Squadron operated exclusively against that line. There have been statements in the press to the effect that the Canton-Hankow railroad declared a $4,000,000.00 profit during the past year while the Japanese were continually bombing it.

To support our belief that the Japanese and the contestants in Spain were demonstrating an extremely low order of bombing accuracy as we know it, we have previously had only the opinions of various observers. There have been published statements from observers in Spain which have related that about 2,000 bombs had to be dropped in order to put a bridge out of commission.

Even in the days of the D-4 bomb sight in the Keystone, under target range conditions, if one hundred 1,100-pound bombs were dropped from 12,000 feet altitude we believed that one was practically certain to hit within fifteen feet of a ten-foot pier of a heavy bridge and cause one or more spans to collapse. We believed that 2,000 bombs were needed in Spain (and a similar number in China) not so much because of the presence of active antiaircraft defense agencies, not so much because bombs too small in size may have been used; but chiefly because the bombardiers have been unable to put their bombs on relatively very small targets presented by sensitive points in rail lines of communications.

The attached chart [unavailable in the archives] is the first authentic statement that has reached this school from which a precise determination of Japanese bombing accuracy can be deduced. Lt. J. D. Shaw, U.S.N., who submits this chart and described the conditions under which these bombs were dropped as "personal observation," is listed as Executive and Navigation Officer on the gunboat *Mindanao* on the South China Patrol. He is probably not a bombardier, but the pains he has taken to prepare a sketch of the objective with all points of impact drawn to scale is certainly prima facie evidence that these data are reliable.

From this chart, the Mean Circular Error of this bombing is 678 feet. Under similar conditions, at 8,000 feet altitude in the annual bombing matches, with D-4 bomb sights and Keystones, the Mean Circular Error was 254 feet. The Japanese error is nearly three times as great as the D-4 error—hence, the Japanese bombing might be called nine times poorer. What M-1, B-18 errors are today is unknown, but we are sure, under similar conditions, that they are less than the 133 foot errors originally obtained with the M-1 in the B-10.

With the degree of accuracy shown by this bombing, the Japanese would have to take at least 227 *shots* for a *90%* assurance of getting *one* hit in a circle 150 feet in diameter.

We believe—if 227 bombs must be dropped to get one in an area bigger than a baseball diamond—it may be better to employ bombardment to augment the artillery than against small vital objectives.

We have proof that the bombardiers in the Army Air Corps can hit small targets.

We believe that United States Army Air Corps bombardment units can be counted upon to hit and destroy small vital hostile objectives, whatever they may be.

We regret that the data upon which these figures are drawn is "Confidential."

CONFIDENTIAL

Copy
INTELLIGENCE REPORT
Serial No. 34-10   File No. 1005-500
U.S.S. *MINDANAO*   Place: Canton, China.   Date: 16 August, 1938.
Name of Intelligence Officer: J. D. SHAW, Lt. USN, Executive and Navigator, U.S.S. *Mindanao*, River Boat, South China Patrol.
Name of C.O.—J. P. CLAY, Lt. Comdr., USN.
Source: Personal Observation.
Subject: Sino-Japanese Conflict AVIATION—OPERATIONS—TARGET PRACTICE

CONFIDENTIAL

The enclosure marked sketch "A" is submitted as an indication of accuracy obtained by Japanese bombers. These bombs were aimed at an aircraft repair and assembly plant located on the bank of the Pearl River in the northeast suburbs of Canton, China.

None of this bombing was in formation. The altitude was about 8,000 feet, flat bombing, medium sized landplanes, with no interference from anti-aircraft fire. The general direction of the flight is as indicated on the sketch. Location of bombs are plotted to scale. All bombs were 120 pounds except those marked A and B which were believed to be 480 pounds. The point of aim should have been quite clear as it was a large building which, though camouflaged, was well set off with trees around it.

It is hoped that these reports will provide some basis for comparison with the results obtained by our bombers under like conditions.

This report is submitted separately from #33-9 as all bombing of the latter report was done in formation.

CONFIDENTIAL

Submitted by J. P. CLAY, Lt-Comdr., USN.
Forwarded by J. P. CLAY, Lt-Comdr., USN.

In this period we have discussed the tremendous increase in the effective strength of the GHQ Air Force that might result in lowering mean probable errors from the vicinity of 150 feet to the vicinity of 50 feet. It was this operation which permitted us to use five flights where ten squadrons had been required before. Here are two selected bombing records of recent bombing with the Mk I bomb sight in the B-18 airplane. In this first scoring record from 11,600 feet altitude by Lt. Offut, pilot, Lt. Schwanbeck, bombardier. They dropped five bombs from the clover leaf method of approach from 11,600 feet altitude in a 15 mile wind from a true air speed of 156 miles per hour. Their mean probable error was 58 feet. In this other record Captain DeRosier and Lt. Anderson, bombardier, in a B-18 airplane, took off from their base en route to the objective, climbed 10,000 feet (indicated 9,400 feet) and bombed target 2,610 feet higher altitude than their point of take-off. They operated in a 25 mile per hour wind with a true air speed of 150 miles per hour. They dropped eight bombs, the mean probable error of which is 38.5 feet. Of their eight, the fourth, fifth and sixth were all considerably short because of unknown conditions. If those three bombs were eliminated the mean probable error for the remaining five bombs was 21.1 ft.

Please do not infer that these two scoring records prove that the errors tabulated in your text and employed in Practical Bombing Probabilities Problem No. 2 can arbitrarily be reduced to about ⅓ their present magnitude. These two particular records were selected to prove that it is *possible* to reduce the errors we have employed to about ⅓ their present magnitude.

These records do prove that it can be done. Such bombing increases the effective strength or the actual power of the 7th Bombardment Group six times.

It is possible—but it can be accomplished only by enthusiastic, interested control and guidance in training.

Most of us have had contact with cases where bombardment commanders have worked units on showy formation flying—almost to the exclusion of other types of training. Many of us know situations where bombardment training has been concentrated upon the ability of bombardment squadrons to pitch up tents in long straight lines—with a wail when higher headquarters interrupted such "training" with unreasonable demands that the bombardment outfit go some place and bomb.

More than one Air Corps officer has been in command of bombardment units or charged with bombardment training who felt that bombing was a disagreeable interruption in the proper training of his bombardment organization. When a Field Artilleryman refers to a good battery, he is speaking of an organization which, among other things, goes into action rapidly and delivers fast accurate fire. When the average Air Corps officer refers to a good Bombardment Squadron, he too frequently refers to an organization which can fly big airplanes in a precise review formation. We have excellent Training Regulations and well prepared training directives. We know that all of the official publications under the sun can never build expert bombardiers in any unit unless there is an interest in accurate bombing.

We have devoted four conferences to bombing accuracy as represented by probable errors which determine the capabilities of Bombardment Aviation.

Today's assignment indicates that our conception of the employment of bombing errors by the bombardment commander is a miniature of some of the Air Corps Board's plan. You may have felt that we were advocating extensive paper work in statistics. In this case we do not hesitate to affirm that we are advocating extensive computations. We feel that the bombardment squadron armament officer and his reserve assistant or two may actually be given an interesting, important and profitable job. The armament officer may be transposed from a fifth wheel on the flying line to an important statistical aid to the bombardment S-3. Today's assignment is believed to be sufficiently complete to stand without elaboration.

We do not expect the bombardment commander to ever determine the single-shot probability of any single target. We do expect that he will insist on having the most accurate means of determining how large a force he must employ in order to attain a reasonable chance of success. He will, therefore, demand that the probable errors of teams, flights, squadrons and groups be accurately determined. He may well have some staff officer or perhaps a mathematically inclined corporal take an objective folder and the latest probability data and compute the probable chance of success of an airplane, flight, squadron and group when operating in each of the various conditions which effect bombing accuracy, from all altitudes from zero to ceiling.

If the commander has that information—if he is confident that that data is accurate—then he has the best scientific aid to determine "how large a force, under what conditions, from what altitudes, will produce a reasonable chance of success."

Comments on "Practical Bombing Probabilities"

This remarkable lecture acknowledged that the Air Corps had severe deficiencies in its ability to determine the bomber force required to destroy any given target. The ACTS set about solving this problem by training airmen how to calculate mean probable errors. This new science would provide the foundation for air force tactics and operational planning for the next six decades.[7]

As fundamental as this work proved to be for determining the feasibility of high-altitude daylight precision bombing, it largely ignored the equally important consideration of how to locate and identify the assigned target. Flying at high altitude increased the likelihood of encountering weather that would obstruct the line of sight from the bomber to the target. While inflight visibility might not be as significant an issue when flying over the dry desert terrain of Kuwait, Iraq, Afghanistan, or Libya, it would prove much more challenging in the humid climates of Germany, Japan, Korea, Vietnam, Bosnia, and Kosovo.

ACTS largely ignored these questions of target identification. Armed with the B-17, its Norden bombsight, and with planners skilled in calculating the bombing probabilities, the next critical question was to determine what targets should be attacked. The next chapter addresses this issue by specifying the vital and vulnerable infrastructure nodes that, if disabled, would paralyze the enemy's wartime economy and wrest from it its capability and/or will to resist.

5

Vital and Vulnerable

NATIONAL ECONOMIC STRUCTURE AND
NEW YORK INDUSTRIAL AREA

By the late 1930s, the combination of the B-17 and the Norden bombsight had finally provided the Air Corps with a technological solution for the conduct of long-range, independent air operations. The next question to consider was what targets should be attacked to effectively and efficiently bring about victory from the air. Donald Wilson first provided the ACTS solution to this targeting problem with his two lectures "National Economic Structure" and "New York Industrial Area." The US foreign policy of isolationism had discouraged offensive war planning against potential adversaries and thereby prevented the Air Corps from obtaining intelligence on Germany or Japan. Wilson instead took a different tack, analyzing the extent to which America's economic infrastructure was most vulnerable to enemy attack. Specifically, he identified where US industrial capacity was most concentrated, the power sources required to operate these factories, and the transportation networks needed to link production.

Whereas in "National Economic Structure" the focus was on how an enemy nation might target the United States, in "New York Industrial Area" the main point was that other modern states, with the exception of Russia, did not have the continental size and dispersed economic structure of the United States. Great Britain, France, Germany, and Japan concentrated their economic and political power in their capitals. The lecture analyzes New York City as an example of how to determine what to target in a metropolis.

Regrettably, only rough outlines of Wilson's original 1936 lectures remain. His successor, Muir Fairchild, however, expanded on Wilson's work with two superb lectures that have been preserved, except for the supporting charts and maps, which were so large (to enable the entire class to see them) that they could not be placed in the archive files. The tables, however, have been preserved because they were typed into the text. Together, these lectures provide the key for identifying the vital and vulnerable nodes of an industrial nation's infrastructure which, when attacked, would result in the economic paralysis that the ACTS believed would be decisive.

National Economic Structure

Major Muir S. Fairchild

April 5, 1939

Gentlemen, in our conference on Air Warfare so far in the course, we have referred many times to the ability of air forces to attack the enemy nation directly.[1] It becomes apparent, then, that the first and basic choice in how air forces are to be employed is whether we are going to choose to do *this*, or whether we are going to confine ourselves to the attack of the hostile armed forces.

Before we can intelligently take up this question it is necessary for us to establish what we mean by the direct attack of the enemy national structure. Also we must try and reach some conclusions as to how we are going to go about it, and just how vulnerable such a structure *is* to air attack. In other words, how effective may we expect such attack to be and how are we to conduct it. If we can reach some reasonable and logical conclusions along these lines we will be in a better position to decide between taking *this* course of action, and choosing the enemy armed forces as our primary objective.

Now when we speak of direct attack of an enemy national structure, I have no doubt that it naturally connotes to you, the direct attack of the civilian population—the gassing and bombing of the concentrated populations in the great cities. Certainly we must admit that this is one recognized method of making such an attack. It is certainly true that most of the European nations are definitely contemplating such a method of attack—if not for their own forces, then surely for the hostile forces that they expect to be directed against them. When we read that Great Britain has produced over 40 million gas masks at the present time—with which it is proposed to equip every man, woman and child of the English population—it is convincing proof that they regard this method of attack as something more than a remote theoretical possibility.

Last year an item appeared in the press describing a debate in Parliament over a bill authorizing the government to remove by compulsion, and distribute over the countryside, some seven million people from the more congested areas, whose presence would not be necessary in those areas in war time. The bill was opposed by one member who pointed out that it would be nearly impossible to move such a horde of people from their usual places of residence and to distribute them to other, less densely populated areas. This opposition was effectually squelched by the government spokesman, who merely remarked that, impossible or not, it would be much easier to remove the people while they could walk than after they had become stretcher cases. The bill thereupon was passed unanimously by the House of Commons.

France and England are both building gas- and bomb-proof shelters on a tremendous scale. In Paris the subways are being utilized in many places, and sections that will accommodate as many as 5,000 persons are being blocked off and rendered gas- and bomb-proof. Building regulations and laws have been placed in effect in these countries to require the inclusion of gas- and bomb-proof chambers in new buildings that are to be erected in the future. You will remember from the Antiaircraft Defense course the extensive preparations for air attacks that have been instituted in Great Britain by the Air Raid Precautions Department.

I mention all these indications because they are proof positive that those nations that are today in a position where they may be subjected to direct attack by Air Warfare, quite evidently feel that at least a considerable portion of that attack will be directed against the civil population.

That such an event is a distinct possibility is not open to question. There is, however, considerable question that this method of attack is the best and most efficient method. This point is recognized to some extent by the European nations referred to and we find them, not only taking all the steps they consider feasible to afford protection to their civilian populations, but also employing means of passive defense to provide the greatest possible protection to other parts of their national structures as well.

Now *is* the direct attack of the civilian population the best method of bringing pressure to bear upon a hostile nation so as to achieve our aim in war by means of Air Warfare? There is no positive answer to this question since it has never been tried as yet against any highly integrated major nation. However, the School does not believe it to be so.

In the first place, the direct attack of civilian populations is most repugnant to our humanitarian principles, and certainly it is a method of warfare that we would adopt only with great reluctance and regret. Furthermore, the repercussions in neutral countries could not be completely ignored, though it is apparent from what has occurred in Ethiopia, Spain and China in this respect, that this is not a matter that need cause us as much concern as was formerly thought to be the case. In those undeclared wars, thousands of civilians including women and children have been killed during the bombing of cities without the reaction in neutral countries being serious enough to have any real effect.

However, there is another point which has a real bearing upon the problem and that is whether this is the most *effective* method of waging air warfare directly against the enemy nation. Obviously we cannot and do not intend to actually kill or injure *all* of the people. Therefore our intention in deciding upon this method of attack must be to so reduce the morale of the enemy civilian population through fear—fear of death or injury for themselves and their loved ones—that they would prefer our terms of peace to continuing the struggle, and would force their government to capitulate. Any pressure that exerted upon the war making capacity of the hostile nation, other than this, would be entirely incidental.

Now the building up of morale during war has always developed from the act of war, and it has required a very appreciable length of time to develop to the maximum. Once developed, however, the breaking down process by the application of military pressure has taken an even greater length of time. Inasmuch as we may expect the initial mental state of a nation at war to be one of confusion, rather than the firm state of morale which may eventually emerge, it may well be possible for air attack direct on the civilian populace to destroy morale before it really comes into existence—provided of course that the air force can strike soon enough and hard enough. The point in doubt here is *how hard is hard enough?* This we don't know; but we do know that air attacks such as the Japanese have directed against the Chinese cities are *not* hard enough, for *there,* the reaction seems to have been to increase the morale of the nation as a whole. Observers seem to be of the opinion that the Japanese air attacks on cities have had the effect of unifying the Chinese people in opposition to the Japanese to a greater extent than any other factor.

It is apparent that if Air Warfare is to succeed in the direct attack against morale, it must create such an abnormal environment that continued sacrifice of the individual for the common good becomes intolerable. It is therefore clear that if this attack is launched against the civilian population directly, and it is to succeed, we must overwhelm the population with fear very quickly. The adaptability of man is very great and observers have commented upon the fact that in the bombing of Madrid and Barcelona, for example, the population eventually became accustomed to the attacks to the point that the fear, which they had at first caused, ceased to be apparent.

In general, the direct attack of populations gives temporary effects only and these are not necessarily cumulative. Furthermore, aside from the psychological effects on the workers, this attack does not directly injure the war making capacity of the nation. For all of these reasons the School advocates an entirely different method of attack. *This method is the attack of the National Economic Structure.*

The School believes that this method of attack is more in keeping with our humanitarian ideals, and more acceptable to those neutral states that might otherwise be influenced against us. Fundamentally, however, this method of attack has the great virtue of reducing the capacity for war of the hostile nation, and of applying pressure to the population both at the same time and with equal efficiency and effectiveness. The results that are achieved by attack of the national economic structure are also cumulative and lasting. They build up from day to day and from week to week so that the pressure that formerly has been imposed by military action over long periods of time may, by this method, be concentrated into a short period, and still produce that intense suffering upon the civil populace that has been essential for the collapse of the national morale and the national will to continue with the war.

Now since this seems to be the best and most effective method by which air forces may apply pressure directly against the enemy nation, let us examine this National Economic Structure and see what we may hope to achieve by attacking it, and how, in general, we are to set about it.

Modern war, as you all know, is dependent—in fact, is absolutely dependent, upon the capacity of the warring nation to turn out the great amount of munitions, supplies and equipment of all kinds required to equip and sustain the armed forces. The World War from start to finish was an economic struggle. The main battlefields were in the industrial areas, and the main weapon was the blockade, in one form or another, which prevented these areas being supplied. You are all familiar with the naval blockade of Germany and the measures that were adopted by the Allies to choke off imports from neutral continental countries. The ultimate effect is well known. The collapse of the

Central Powers was internal and was not caused by defeat on the Western Front but by defeat on the Home Front.

Some idea of the growth in consumption of munitions in modern war is given by the following figures quoted by General Fuller: "For the preliminary bombardments of the Battle of Hooge we fired 18,000 shells; for those at the Battle of the Somme 2,000,000 shells; and for those at Ypres the same year, 4,300,000 shells. At the last mentioned battle the tonnage of shells fired during the preliminary bombardment alone, amounted to 107,000 tons, the cost of which has been estimated at 22,000,000 British pounds (approximately $110,000,000) a figure very nearly equal to the total yearly cost of the pre-war British Home Army."[2]

Confronted with such figures which relate to the consumption of only one item of munitions, as a preliminary to one battle, it is easy to understand that the production of the munitions required to sustain modern military forces strains the industrial capacity of any nation to the utmost. Tremendous sources of raw materials—all raw materials—must be available; enormous facilities for production—for fabrication and processing, must be adapted to the manufacture of weapons, airplanes, ammunition, and the astronomical quantities of all the other munitions of war that are consumed unceasingly by the military forces.

Thus, the nation that mobilizes her manpower must be able to equip it —must be able to provide the modern implements of war required in such enormous quantities—and must be able to maintain the constant flow of supplies and replacements, if that force is to be able to act. It is plain that a nation must possess a highly organized and smoothly functioning economic system to carry on war in the modern way. The capacity to wage modern war is definitely fixed by the capacity of the national economic structure to provide the raw materials and convert these materials into the sinews of war.

This basic fact was thoroughly impressed upon all belligerent nations in the World War. As a result of the disastrous lessons that *we* learned from our own industrial effort in that struggle, our War Department places the greatest stress upon the fundamental importance of the economic effort for future war. To that end we have established a complete planning agency in the Office of the Assistant Secretary of War which is constantly engaged in the full time job of planning our industrial war effort, to the end that our whole economic and industrial system can be thrown into high gear and placed under the most extreme pressure, immediately upon the outbreak of hostilities.

In spite of the fact that the United States is the greatest industrial nation in the world, and in spite of the fact that every effort was made to get our great industrial machine to work smoothly and efficiently in the World War, our record in that respect is not too happy. It was found that the capacity of our industries, which is frequently taxed during normal times to supply the peacetime demands, was seriously strained when it was required to take on the additional demands of the military forces. It had to speed up to the limit to keep pace with the enormous demands made upon it. In this mere process of speeding up, all sorts of dislocations occurred. This economic disarrangement brought about, among other things, rapidly rising prices, food and fuel shortages, transportation congestion, labor unrest, and suffering and weakened morale among a large portion of the civilian population. At one time every siding and much of the rail line was filled up, from the port of New York as far west as Pittsburgh, with backed up freight trains, caused by congestion due to the war load. This condition existed for almost three months, during which time trains could be moved only with the greatest difficulty, if at all. The repercussions were far-reaching and seriously retarded our war effort.

Innumerable examples could be cited to show the difficulties which this great industrial nation experienced in attempting to meet the war-time load that was placed upon industry, but time is not available for this historical examination. I think, however, it is quite plain that modern warfare places an enormous load upon the economic system of a nation, which increases its sensitivity to attack manifold. Certainly a breakdown in any part of this complex interlocked organization must seriously influence the conduct of war by that nation, and greatly interfere with the social welfare and morale of its nationals. No nation at the outbreak of war has a sufficient supply of war material on hand. It must depend on a program of rapid expansion that will increase production and distribution to satisfy requirements.

It is characteristic of modern civilization that the economic structure is dependent as a whole upon the integrity and continued functioning of each one of its individual elements. Raw materials must be available. They must be transported to the processing and manufacturing centers and from there moved on to numberless other assembling and finishing plants; through a long chain of operations, with many stops on the way, before the completed article is ready for final delivery. And everywhere there must be available—POWER. Without power the whole vast machine comes to a jarring halt.

This system is highly sensitive in peace. A strike in a small obscure factory producing door latches for automobiles stopped production in many automobile factories all over the United States. A spring flood in Pittsburgh stopped the production of airplanes in California as well as in other sections of the country. If this is true of our industrial machine in peace, when it is under minimum load, if it is this sensitive then, how sensitive will it be during war when it is under maximum pressure?

The application of the additional pressure necessary to cause a breakdown—a collapse—of this industrial machine by the destruction of some vital link or links in the chain that ties it together, constitutes one of the primary, basic objectives of an air force—in fact, it is the opinion of the School that this is the maximum contribution of which an air force is capable towards the attainment of the ultimate aim in war.

If there is any one point that the study of air attack on the National Economic Structure impresses upon us, outstandingly, it is that the objectives to be attacked in this type of an air offensive are not to be selected on the morning of the attack. The future general of our air forces will not, I hope, be faced with the decision that must select either bridges or refineries for the next air mission. Let us hope that war does not find him bent over a map in his tent, trying to pick out the vulnerable points to be hit by his attacking force. The picture is to be quite the contrary, I think.

Complete information concerning the targets that comprise this objective is available and should be gathered during peace. Only by a careful analysis—by a painstaking investigation, will it be possible to select the line of action that will most efficiently and effectively accomplish our purpose, and provide the correct employment of the air force during war. It is a study for the economist—the statistician—the technical expert—rather than for the soldier.

War plans for ground forces cover in great detail the successive steps of mobilization and concentration of the forces to be involved. They may indicate in general the plan of campaign and the initial objectives of those forces. On the other hand, a war plan covering the employment of air forces where fixed objectives are involved, as in the attack of the national economic structure, should be a detailed plan of attack—a plan of actual operations.

Such war plans may well have been prepared for the operation of some of Europe's great air forces. There, definite and well recognized enemies, with highly organized economic structures, lie completely within the tactical radius of action of opposing air forces. The respective economic structures, with their vital and most vulnerable links, are well known to enemy nations.

In this country none of these conditions exist at present and such war plans, of course, have not been drawn.

In order to get some idea of what such a war plan would be like and how we should go about preparing it, let us devote the time this morning to attempting an analysis of a national economic structure. Lack of time and talent necessarily will result in a very sketchy analysis at best. Therefore let us choose for this analysis that country with which we are all most familiar—the United States—and try and place ourselves in the shoes of the enemy that contemplates bringing the attack against us.

With a subject as broad as this we could spend a great deal of time, and still not arrive at the final answer. Consequently we will not try and reach definite conclusions, but will merely hit a few of the high spots and see what we get out of it. If we should be able to put a finger on any weaknesses in this industrial giant of ours—whose sources of raw materials are unsurpassed—whose production facilities are unequaled—we can understand better what the possibilities of air attack may mean to less fortunate nations who have not been so bountifully endowed with resources and facilities.

Let us take up, first of all, a very brief consideration of our geographical advantages in relation to the other great powers.

It must be realized that except for Russia, this country is the only industrialized power built on continental proportions. We enjoy, therefore, unique super-abundance in the foodstuffs and raw materials essential for the development of national power—both in peace and war. But even in the case of those commodities, which we would be unable to produce in sufficient quantities under a war emergency, our situation is likewise unusual. Compare our position with that of other great powers, including their colonial possessions as shown on the map. The most notable feature at once apparent is the large proportion of the earth's surface contained within the jurisdiction of the United States, Russia, and the empires of Great Britain and France. The vast *consolidated* territories of the United States and Russia furthermore make a striking contrast with the scattered and, therefore, strategically more vulnerable possessions of the other great powers. Consider the position of the United States insofar as the availability of imports of raw materials in time of war is concerned. Contrast our situation with that of the other great powers. Russia, for example, is hemmed in by both Asia and Europe with no direct outlet to the sea. As to Great Britain and France, imports of necessary supplies from their colonial possessions, or other extra European regions, are extremely vulnerable in time of war. They must eventually be transported through the dangerous bottleneck of their territorial waters, near which lie the centers of their industries. In the case of Japan her complete isolation from the other industrialized centers of the world renders her situation precarious. The lack of a large number of important raw materials, which she and her immediate neighbors suffer, definitely limits the size and self-sufficiency of her industry. In any major conflict she must resort to western industrial powers to provide a considerable number of war commodities.

By way of contrast, the geographic situation of the United States is such that none of these disabilities of the other powers apply, to anything like the same extent. Our direct access to the two oceans of world trade releases us from the same dangers of the bottleneck. So much for a comparison of the geographic aspects of the great powers.

Let us consider now their relative self-sufficiency in raw materials as compared to the United States.

[Here and throughout this section, the speaker refers to the chart of strategic materials, which is not included in the archive.]

This chart is designed to indicate the strategic situation of the great powers in terms of relative self-sufficiency in foodstuffs, essential industrial products and raw materials. It also indicates the comparative normal peacetime consumption of the several great powers in each of these items.

The chart has two scales—a vertical scale and a horizontal scale. The solid black portion of the vertical scale indicates the percentage of the apparent national annual consumption of each particular item which is derived from domestic production. The hatched portion of the vertical scale indicates the percentage of the apparent national annual consumption of the particular item which is obtained by imports from abroad.

For example, we see that in the case of tungsten, the United States produces about one-quarter of the amount we consume (as indicated by the solid black portion of the vertical scale). The remaining three-quarters of our annual consumption we import from abroad (as indicated by the hatched portion of the vertical scale). The round dots at the top of the vertical scales indicate that domestic production of that item is not only sufficient to meet all home consumption, but is sufficient to provide a considerable surplus for export.

The horizontal scale indicates the apparent national annual consumption of the particular item expressed as a percentage of the annual consumption of the nation consuming the greatest amount annually of that particular product.

For example, the United States consumes the most *food* and hence on the horizontal scale it rates as 100% covering the whole of the horizontal space. In comparison, Germany and Great Britain consume only 25% as much food and hence are shown on the horizontal scale as covering only one-quarter of the horizontal space.

You will note from this chart that a comparison of the normal peacetime situations of the seven great industrial powers reveals some striking contrasts. One of the most striking facts illustrated by the chart is, of course, the great degree of dependence of the other powers upon imports of raw materials. It will be observed that the degree of self-sufficiency of the United States in the great essentials is very superior, Russia alone being able to approach our advantage in this respect.

In respect to consumption, it will be noted that the impressive position of the United States becomes at once apparent, for with the exception of nitrates, potash and wool the demands of our industry are not only larger than those of any other nation but for the most part are equivalent to the total demands of the other powers combined. Viewed from this angle the apparent self-sufficiency of a nation like Japan in coal, chromite or tungsten, or Russia in coal and iron, is not particularly impressive if applied to the normal consumption capacity of American industry.

Note the factor of power shown in the second column. This factor is of conclusive importance in industry. Please note particularly the dominating position of the United States in this respect. Note also that in power we are completely self-sufficient, since this power is derived from three sources—water power, which, of course, is our own—coal and petroleum, of which we not only have a great sufficiency but an exportable surplus. Note the position of the other great nations in this respect.

The United States, even with such a generous supply of raw materials within its borders, still imports from all over the world other raw materials it needs in its manufacturing operations, and pays for these materials with finished articles. With about seven percent of the people of the world we do about one-half of the world's work. We can do this because our daily output of power is about 13.5 horsepower hours per capita, as compared, for example, with less than one for Brazil and Peru. Another factor which enables us to accomplish so much is that about 22% of the people of the Nation can raise the food and vegetable fibers to feed and clothe the whole Nation; thus leaving 78% free to take part in commerce or industry, or to fight a war, if necessary. This 22% compares with about 40% to feed Germany, 45% for France, 65% for Italy, and 80% for Russia, which must be "on the land." Due to the horsepower helping an American workman, and to his own training and skill, he is equal to two Frenchmen, or nearly two Germans, or three Italian workmen, or three and one-half Russians, as measured by the actual work accomplished. These conditions cause the United States to use about one-half of the raw materials of the world.

Although this nation is the most self-contained of all nations, there are many materials we do not have, and these we must import, if our production, our war effectiveness, and our high standard of living are to continue.

Therefore let us briefly consider some of the materials we do not have, or do not have in sufficient quantities—strategic raw materials. Strategic raw materials are those for which dependence must be placed on importation from abroad, and of which domestic production can supply only a small part. They are twenty-six in number, and include such materials as ferrograde manganese, tin, chromium, tungsten, and wool.

Just to give an idea of what constitutes a strategic raw material—where it comes from and what it is used for—let's very briefly examine one or two of them.

MANGANESE ore, for use in making steel, is normally imported. Our country can produce low grade ore and chemical ore but has no large deposits of the richer ores. It is impossible to make sound steel without this high grade ore, and without sound steel our industries and our war efforts would be paralyzed. The world's supply of high grade manganese ores is found in Russia, India, the Gold Coast of Africa and Brazil. All of these courses are distant from our steel industry and ocean shipping is required to bring this essential material to us. It appears that nothing short of a physical reserve of this material will make us completely safe from interruption and a bill is now before the Congress to authorize the expenditure of some millions of dollars in building up stock piles of this and similar materials.

TIN is one of the minerals which nature left out of our soil. It has become indispensable in the preservation of food, since it furnishes a protective coating on the steel with which our so-called tin cans are made. We use it in the manufacture of automobiles, for the making of bearings, solders, bronzes and gun metals. The world's sources are the Straits settlements and Bolivia.

RUBBER grows in the tropics, and we are dependent upon outside sources for our supply. We use over one-half the world's production, which comes from tropical countries distant from our shores. We have developed several substitutes but our production capacity for these is small and the cost of producing them is higher than is the natural crude rubber. At present these substitutes are being used by industry only in the special cases where they are better than crude rubber for some purposes. Without the imports to which we are accustomed our industry and our people would suffer serious hardships.

TUNGSTEN is used as an ingredient of tool steel, since this material in steel enables a cutting tool to keep its edge even when hot. For production during an emergency, when speed is essential, it is hard to see how we could keep the pace required without this valuable mineral. Smaller quantities of this material are used in steel which has to stand excessive heat. The world's major supply comes from China, with lesser amounts from India and Bolivia.

Practically the same conditions exist in regard to our other strategic raw materials: nickel, chromium, antimony, camphor, hides, mercury, mica, nitrates, platinum, coconut shells, silk, wool, shellac, food items such as sugar and coffee, vegetable fibers for rope, bagging and twine, and certain essential drugs, such as iodine, quinine, and the opium derivatives.

Now it is apparent from this brief analysis, that even with by far the most favorable situation economically, among all the great powers, we could still be seriously embarrassed by the denial of certain strategic imports.

As was brought out by our analysis of our geographic situation, however, we are very favorably situated. We have access to the two oceans of world trade. All of our strategic materials come from at least two, and in most cases from many, sources at widely scattered places on the earth's surface. In our case they do not have to pass through the narrow bottlenecks which we noted existed in the case of Great Britain and France. They may be unloaded at any one of our many widely scattered seaports.

We are forced to the conclusion that our prospective attacker could hardly hope to gain any decisive results by employing his air power against this external element of our national economic structure. It is simply too great a task. Too many specific objectives are involved and they are too widely dispersed. The task would almost certainly be beyond the capacity of his available force.

However, let us recall the geographic situations of the other great powers and note their degree of dependence upon imports of essential materials as shown by this chart. Note that Great Britain is dependent upon imports for over half of her food supplies. Those vital imports of food must be brought through the narrow bottleneck of her territorial waters to only a few ports, all fairly concentrated geographically. Again, Germany and Italy are among the pauper nations insofar as raw materials are concerned. Japan is but little better off.

Considering these facts we must conclude that a careful detailed analysis of the economic structure of some of the other great powers would probably give a far different answer in this respect than in the case of the United States. Such an analysis might very possibly come out with the answer that in some cases the denial of imports of essential materials would constitute the vital and most vulnerable element—one that was well within the capacity of the force and which would be quickly and conclusively decisive.

We will resume our analysis of the economic structure of this outstanding industrial country—the United States—after the usual intermission and we will continue our efforts to find the vital and vulnerable link which surely must exist somewhere in this great economic structure.

Having sketchily surveyed the economic factors involved in the geographic and external situation of the United States, let us turn our attention to the internal economic structure of the country. We have noted the fact that the United States leads the world in industrial capacity. We will take a quick look at a few of the outstanding factors that go to make up that great industrial capacity and see if we can locate some link in the chain which is at the same time both vital and vulnerable to air attack.

One of the most important industrial factors in the United States is the industry concerned with the production and the refining of petroleum. Petroleum is essential to modern war. Air forces, and mechanized and motorized forces, would be helpless without its products. Should it be possible to deny oil to a nation, its ability to wage modern war would be seriously interfered with, if not completely terminated. Also this denial would cause a breakdown in the transportation and distributing system of the nation, which is dependent, at point of origin and destination, upon motor truck operation. Such a breakdown would impose tremendous pressure upon the civil population, especially in the great cities.

The United States leads the world in the production of petroleum and its refined products with 60% of the total. It also leads the world in its consumption. We produce more than one billion barrels a year from our fields. The petroleum industry is a vital industry to the United States and it is worth the while of the potential attacker to study it and decide if he has the capacity to destroy it.

[In this section the lecturer refers to a map of oil fields, no longer available.]

Note the wide distribution of the oil fields themselves. Note the pipe lines that lace the country. There are a total of 58,000 miles of main crude oil lines; 54,000 miles of feeder lines; 50,000 miles of gas lines; and 4,000 miles of gasoline lines. Note the location of oil fields near the Atlantic and Gulf Coasts and the ready availability of water transport for both the crude and its products. In addition, an excellent rail net interlaces all of these fields. The facilities for movement by rail are enormous: more than one-half of all refined products moving by rail and between 6 and 7% of the crude. Existing pipe lines have a capacity sufficient to move a high percentage of the crude, though their routes are, of course, fixed. 25% of the total tonnage in American registry are tankers. One-third of all Atlantic coastwise shipping is petroleum or its products; two-thirds at the Gulf ports, and one-half at Pacific ports. Note the concentration of refinery facilities, represented by the black dots. The northeast vital area consumes 55% of all gasoline produced in the United States and produces 36% of the total, or 70% of its own supply. It has 113 refineries but 50 of these produce 91% of the total, and 33 refineries produce 80%, all of these being concentrated under the black dots.

Now there are not a great number of these black dots, and refineries being as vulnerable to air attack as they are, it should be possible to destroy these 33 concentrated refineries. This would certainly be a most serious blow to our vital area—the source of over half of its gasoline supply would be wiped out. This would certainly apply a great deal of pressure and it would cause all sorts of dislocations in industry and the normal life of the people. But would it be conclusive?

We have noted that petroleum and its products are very easy to move and that ample means of transportation exist. So long as the distribution system remains intact and since the capacity of the remaining refineries of the country is sufficient to meet the worst of the shortage we can still move in the bulk of the requirements of the northeast.

From this very brief and sketchy analysis of this great industry we are forced to the conclusion that, as potential attackers, we had better look further for a more vulnerable section of the economic structure if we are to apply *conclusive* pressure against the United States.

However, perhaps we should not leave this industry, as students, without reminding ourselves that the United States is unique among the nations of all the world in respect to petroleum. Surely a comprehensive similar analysis of most of the other great powers, which produce either no petroleum at all, or but a minor portion of their requirements; who have no such pipe lines or other distribution systems and whose storage facilities are *concentrated* rather than dispersed—would tell a very different story.

Having failed to find that the great national petroleum industry constitutes a very vulnerable link in the national economic structure of the United States, let us turn our attention very briefly to the *steel* industry. I do not need to stress the fact that if the steel industry can be destroyed, our war capacity is reduced directly in proportion to the destruction achieved. In addition there would result the tremendous hardships which would soon thereafter be imposed upon the civil population.

[Here the speaker refers to a map showing various industrial resources, as he discusses.]

Note the iron ore resources of the United States, shown in orange; the coal resources necessary for smelting, shown in green; the concentration of pig iron production centers, shown as black dots; the movement of ore through the Soo Canal shown in red; and the concentration of steel plants, shown in blue.

What are our vulnerable points here? Certainly the destruction of the Soo Canal locks would be very troublesome. It is a concentrated vulnerable objective. You will recall that during the Bombardment Course on Bombing Probabilities you calculated the size of the force required to give a reasonable assurance of destruction of these locks. Recalling those figures we can conclude that it could be destroyed by quite a reasonably sized force without much difficulty. This would cause the diversion of the 50 million tons of ore, that normally pass through the locks annually, to the lake railroads for an average haul of 864 miles, which equals 43 billion, 200 million ton miles, or a 50% increase of traffic density on all lake railroads. It appears from study that this increase perhaps could be absorbed except for a serious shortage in locomotives. The general dislocation would, however, be very serious and would certainly cut our production of steel products down materially and would be a source of grave disturbance to our economic effort.

It would, of course, be possible to attack the railroads in this connection, but from our very brief and hasty survey it seems more logical to conclude that the real effect would be gained by operations against the concentrated steel mills themselves. Note their very considerable degree of concentration. Remember the historical summary of the effect of the trifling zeppelin raids upon Great Britain's steel production, where the bombs did no damage to steel mills and yet production dropped off one-sixth. Remember what the distracted representative of Messers. Palmer had to say about the effects of the raids and the probable damage that would result to the steel mills if the raids were continued—and that did not include damage to be expected from bombs.[3]

I don't want to draw any definite conclusions here about the vulnerability of the United States steel industry to air attack. It would require a much more detailed analysis to justify definite conclusions. Nevertheless I think it is safe to conclude that it is sufficiently concentrated so that certainly *very considerable* pressure could be applied in this manner. That is the only point I would like to make—that, and the fact, which I hope is apparent to you, that all the necessary information to make the required analysis is available in time of peace—available to all the world. Proper analysis of that information will give us a very definite answer as to the degree of vulnerability and the effect to be anticipated from various degrees of destruction.

Let us now take a quick look at this vital industrial area of ours which has been mentioned several times.

[The "vital industrial area" described is the area between Boston, Massachusetts, Baltimore, Maryland, and Buffalo, New York. The speaker also refers to maps illustrating the population, concentration of industry, and distribution of war load in the area, which are lost.]

Although this vital industrial area of ours includes only 13% of the total area of the United States, it contains only slightly less than one-half of the entire population, and a little more than one-half of all those gainfully employed.

The relative concentration of population in the country is well portrayed by the grouping of the black dots on the map. Of all the persons throughout the United States employed in the occupations shown in table 5.1, the percentages given indicate the relation of this vital area to the whole nation.

Table 5.1.	
Extraction of minerals	47%
Manufacturing	64%
Transportation & Communication	54%
Trade	55%

This area produces 58% of the total United States wealth from forests, farms, mines, and factories. In value added by manufacture, however, it produces some 75% of the entire total. Thus in 13% of the area of the United States, we find considerably more than one-half of the entire economic system of the nation, including the great financial centers, and the political seat of the Government.

The Boston-Baltimore area, which is just a small fraction of the 13%, has percentages of the total war time requirements of the War Department allocated to it under the current Industrial Mobilization Plan as shown in table 5.2.

Table 5.2.	
Cartridge cases	28%
Armor plate	49%
TNT	60%
Inspection gauges	86%
Electrolytic copper for fuses, bands, etc.	69%
Small arms	70%
Nitric acid	52%
Fire control instruments	87%

And so forth for other items too numerous to mention. In fact, out of the total allocations of facilities to produce our entire critical list of requirements amounting to 11,842 factories—three states alone: New York, Pennsylvania, and Massachusetts, have 5,791, or just about one-half.

Here then, is in truth a concentrated objective which one might not suspect existed in this great continental industrialized nation of ours. Had we the time we might make an analysis of this area and the vital industries that are concentrated there. All the information is available for this purpose in various commercial publications—available to us and to our prospective attacker as well. We haven't time to do this but even without it I think we can draw the obvious conclusions that this vital area of ours must be protected at all costs if our economic structure is to continue functioning, and, furthermore, that in this area an adequate analysis would provide our attacker with vulnerable, concentrated targets whose destruction would apply tremendous pressure to our civilian population while at the same time seriously impairing our ability and capacity to wage war.

Particularly we should analyze the transportation system within this area. I am sure everyone will admit that a breakdown of the transportation system would cause the stoppage or complete collapse of all industry within the area. It seems likely from what we know of transportation systems generally, that an adequate analysis would reveal concentrated vulnerable points whose destruction would go far toward causing that breakdown.

It would require a rather comprehensive analysis of available transportation to reach any indication as to whether a reasonably sized air force has the capacity to cause a sufficient breakdown of transportation within this area to accomplish our purpose. While this is a fairly large subject, all the necessary information is available and, time permitting, it is quite possible to arrive at a definite answer.

However, time does not allow us to go into *this* question and still take up an even more interesting, and perhaps important aspect of the national economic structure—the question of POWER.

So as a final item in our very hasty and sketchy survey of the economic structure of the United States, let us now consider very briefly the fundamental item of electric power.

Modern civilization is characterized by the general and increasing use of power and as our industries are now developed they are, perhaps, even more dependent on power than on men or specific materials. The physical limitations on the transmission of mechanical power restricts its use to individual plants, making it of minor importance as compared with electric power. Because electric power is generated in large central plants and it can be distributed over a large area, it has great advantages over mechanical power in cost and efficiency of operation. For these reasons electric power has now very largely supplanted direct connected mechanical power. Therefore, we need to consider only commercial electric power generated by the major inter-connected groups of public service companies. This is the power which must be depended upon to furnish practically all of the processing power to convert basic materials into the finished articles required for war and to maintain our civilian population. According to the 1936 report of the Federal Power Commission, 55% of the total power used in all industrial manufacturing processes throughout the United States is furnished by central public service electric companies.

Many establishments of the primary industries engaging in the initial production of basic commodities, such as metal billets, for example, have their own independent power facilities. However, the present trend for all processing, finishing and assembling plants is to use central commercial power. Furthermore, any additional power required for industry in an emergency must be supplied by the commercial power industry, since there will be no time to construct new power plants.

Now the United States is the greatest producer and consumer of electric power in the world.

[Here the lecturer refers to a Federal Power Commission map, now lost.]

This map of the Federal Power Commission, showing all the important existing facilities as of 1935, vividly demonstrates the importance of electric power to our industry. The size of the power plants is indicated by the size of the circles and squares. The importance of the transmission lines by the width of the red lines on the map. It certainly would appear from the apparent congestion of plants and transmission lines that there was no prospect of a shortage in power in this country; and that it is so wide-spread as to compare with the petroleum industry in point of vulnerability. However, let us go into this matter a little further. I quote from the Interim Report on the National Power Survey, published in 1935, by the Federal Power Commission:

> Very little new generating capacity has been constructed by the privately owned utilities since 1930. As a result the capacity of existing plants is 2,325,000 kilowatts less than the demand that will exist for power upon a resumption of pre-depression industrial activity, assuming maintenance of normal reserve capacity. The critical shortage of existing generating capacity most seriously affects the great industrial districts of the East and the Middle West. It would, therefore, be disastrous in case the United States should become involved in war. The situation might be even more acute than that which ex-

isted during the World War, when in many districts electric service had to be denied to domestic and commercial consumers and nonessential industries to meet war needs for power.[4]

These findings were again confirmed, in the fullest degree, by a government survey of last summer.

[The map showing potential power shortages is not available in the archive.]

This potential shortage is vividly portrayed on this map of the Federal Power Commission. Note particularly the black areas which indicate a potential deficit of over 10%. Please note that this deficit is predicated upon normal peacetime industrial demands. It does not include the additional demands that would be imposed by the war load on industry in a major emergency. Of course, it has no reference at all to any interruption in service, or destruction of facilities, that might be effected by enemy action. It is apparent then that there exists no margin of reserve which it would be necessary to remove by destruction of existing facilities before the effects would be felt by industry. The destruction of even one power plant would be immediately felt as a shortage of power for essential industry in its area.

But perhaps it would be possible to utilize power from other areas to equalize this load. However, the transmission of electric power is always accompanied by line losses which vary directly with the voltage and the distance.

Most large modern plants generate alternating current at a voltage varying from 6,600 to 13,200 volts at the generator, but as alternating current can be easily stepped up or down, the voltage is stepped up by means of transformers as soon as it leaves the generator, sometimes to as high as 275,000 volts. Because of the danger of utilizing power at high voltages the power must again pass through step-down transformers before reaching the consumer.

Even with these extremely high voltages there is a definite limit to the distance that electric energy can be transmitted effectively—the normal limit being from 200 to 300 miles. We see then that we cannot count upon utilizing any surplus power from other districts very far removed from the area where shortage exists.

Furthermore, I would like to call your attention on the Federal Power Commission Facility Map to the fact that there are very few main inter-connections between districts. The light lines shown on the map represent low voltage lines of definitely limited capacity. It is not possible to increase that capacity without re-building the line, and especially, without re-building the transformer stations—a process which would require many months. One of our public utility representatives quoted as an outstanding achievement, the building of 65 miles of transmission line in seven months. No doubt this time could be reduced under war time pressure, but obviously it will always be a time-consuming process. In order to lend authority to the estimated time of replacement I quote briefly from an article by Major General Markham, when he was Chief of the Engineer Corps.[5] Please bear in mind that he is concerned solely with the potential shortage for war time procurement and not with the effects of an enemy air offensive:

> In attempting to provide additional power when a shortage develops in an emergency, the following must be borne in mind: (a) Since the construction of transmission lines is slow and expensive and a transmission of power for long distances is accompanied by large line losses, construction of expensive new transmission systems in time of war should not be counted on. The most practical method of providing additional power in a given area is by constructing new generating capacity in the area. (b) Since the construction of a large hydroelectric plant will require from four to five years the main reliance for additional power must be placed on new fuel plants. (c) Unless prior plans and housing facilities exist new steam generating equipment cannot be installed in less than eighteen months. If such prior arrangements have been made a limited amount of equipment could be built and installed in about nine months. (d) The capacity of the manufacturing companies capable of producing suitable boilers, turbines, generators and other heavy equipment is decidedly limited.

In conclusion I wish to emphasize the importance of power in war time. Our past experiences and our annual surveys have shown that we may expect difficulties and shortages of power in various areas in the event of a major emergency in the future. We must not only continue our present planning activities but we must improve on them in every way possible.

Now let us turn back to our vital industrial area. Study of the Federal Power Commission Chart shows there are roughly forty-nine power plants of 50,000 kilowatts or greater capacity in this area. Some of these individual power plants are very large indeed. Individual power plants exist in the United States capable of producing and distributing as much as a million, eight hundred thousand, horsepower apiece. While there are many more smaller plants they are presumably of importance only to their immediate locality. From what we have dug up so far we may definitely conclude that if these forty-nine plants were destroyed the effect upon the economic and industrial system of the nation would be simply appalling. With no power in the Northeast, or rather with the tremendous deficit in power which the destruction of these plants would cause, our capacity to produce would be completely crippled.

Let us put it another way. Instead of being the most powerful and mighty industrial nation in the world, relying upon our tremendous resources and facilities to provide us with the means to wage a victorious war, we would be reduced to the status of a second, or perhaps a third class power for a period of a year, or perhaps two. Let us note in passing that the very facilities that we would rely upon to repair this damage themselves lie within this area, and themselves depend upon the continuation of power to manufacture the replacement equipment required.

Now what about the vulnerability of this great industry to air attack? If we examine this question we find that it is a very different sort of industry from the petroleum industry. Electric energy must move through a definite circuit, including many highly specialized units all built to handle a specific load. The capacity of these units is definitely limited. The destruction of any one unit in the chain from power source to distribution point will definitely put that source out of action. It is impossible to divert the flow around the damaged unit—such as the step-up or step-down transformers, the turbines, generators, rotary converters or even the boilers of steam plants. It is impossible to store electricity as we store petroleum. Immediately upon the breakdown of any link in the chain the industry served by that system ceases to operate. You have all seen the open transformer units located at either end of large transmission lines. They occupy considerable area and are very vulnerable to attack with demolition bombs, especially in view of the additional destruction that would occur, due to shorts and the burning or fusing of installations requiring months to repair or replace.

The power plants themselves must certainly be considered a most vulnerable target, due to the high speed rotating mechanism, turbines and generators, or the high pressure steam installations. Any detonation that would move the foundations of this machinery or would throw it out of line, would certainly cause its destruction or serious damage.

Now is sufficient destruction within the capacity of a reasonably sized force? Study discloses that a majority of the steam capacity in this area is located in about eight plants. But suppose we include all 49 of the important plants in our estimate. Since you have just finished the Bombardment Course, you are all experts in the theory of probability, so I will leave to you the calculation of the size force that would be required to do this job. That is, unless you would prefer to accept the general conclusion that a very reasonably sized force should be able to knock out practically all of these forty-nine sources in a very short period of concentrated operations. As has already been said, the results would be immediate, cumulative and comparatively permanent.

I would like to call your attention, parenthetically, to the fact that, when air forces are properly employed, their lack of continuous operation or continuous rate of fire, which is so often referred to by the military traditionalist, hardly appears to be the serious limitation which he seeks to make out.

Each mass attack of the power industry that is launched against these fixed, known installations would certainly succeed in achieving some percentage of destruction. Destruction of power plants themselves, thus removing them from the production field; or destruction of the mass of high voltage transformers at either end of their transmission lines, thus preventing the distribution of the power produced.

We have heard the opinion of the Chief of Engineers that replacement would require from 9 to 18 months. We have seen the deficit in normal peacetime production of power for this area, predicted by the Federal Power Commission. We have seen the concentration of industry in this area as established by the Bureau of the Census. We know that one-half of all our allocations of war time requirements are concentrated in only three states of this area. We know that at the impact of each destructive bomb, the wheels of all industry in the area served by that power plant, or that distributing station, stop instantly.

And what of defense against this attack? The locations of targets are well known and quite incapable of removal. They are difficult or even impossible to conceal. In fact, adequate protection by any passive means seems almost out of the question.

As for active defense—the targets are scattered widely over a tremendous area. How would we dispose our six available regiments of AA artillery to offer protection, if indeed, political pressure does not place them elsewhere? And what of our three groups of Pursuit? Where are they to be based to unfailingly interpose themselves between the attacker and his unpredictable target?

What a picture of chaos and confusion all this presents! What happens to our capacity to wage war? Under such circumstances, what is the amount of pressure that would be applied to our civil populations? Would it be sufficient to cause our capitulation before the threat of continued action?

We don't know the answer to this question because such an attack has never been made against a great industrialized nation, as yet. However, if we picture section after section of our great industrial system ceasing to produce all those numberless articles which are essential to life as we know it, we can form an idea of the pressure that would be exerted.

We should note also that the effects are not confined to this area alone. From this area come innumerable articles essential to the continued functioning of industries of every kind in every corner of this country. The stoppage of the small obscure factory producing automobile door latches halted operations in automobile factories all over the country. The denial of a major portion of the production of our vital area, would prostrate every important industry in the United States.

What of the widening circle of repercussions throughout the nation as a whole? While we are unable to evaluate accurately the amount of pressure that would be produced on the national structure, it seems certain that it would be sufficient to strain it to the breaking point, if it did not, indeed, prove to be conclusive.

Now to go back to the purpose we set out to accomplish in this examination of our economic structure. This study was undertaken to see if we could put a finger on any weakness in the economic system of this greatest of all industrial nations. This survey has barely scratched the surface because of time limitations. However, we have demonstrated that ample material is available to anyone who wishes to undertake a real and comprehensive survey. We also have demonstrated, I think, that there are very definite weaknesses in our industrial set-up, and that a properly directed air offensive against our economic structure offers a good chance of complete success in applying extreme pressure against our civilian population—while, at the same time, preventing us from building up, or even maintaining, our capacity to wage war.

Now this discussion would have been much more interesting to all of us, if we had been examining the economic structure of some potential enemy. Perhaps at some time in the future we will have available complete staff studies to form the basis of such an examination. I hope that the discussion this morning has shown their necessity—the necessity of a complete thoroughgoing analysis of this type—if we are ever to consider offensive action against any nation in the future. While the analysis that has been given has only scratched the surface and has been a very amateurish effort, at best, I hope it has demonstrated that a complete analysis by competent personnel is quite possible from commercial sources of information. Thorough analysis in *any* industrial nation will reveal such vulnerability to air attack, as to fully warrant the conclusion that it is in this employment that an air force can offer its maximum contribution toward the achievement of the ultimate aim in war.

Are there any questions or comments?

New York Industrial Area
Major Muir S. Fairchild
April 6, 1939

Gentlemen, the conference this morning is in the nature of an extension of yesterday's discussion.[6] You will recall that yesterday we made a very sketchy and hasty analysis of the vulnerability to a planned air offensive of the industrial and economic structure of the United States. Our purpose in making this analysis was to find out how we could best go about the direct attack of a great industrial nation. We used the United States as an example. Of course it is obvious from the examination we made of our economic structure, that the United States is hardly to be considered a *typical* great nation. In fact, each nation differs from all other nations, not only in its *degree* of vulnerability to air attack, but also in the *kind* of vulnerability; that is to say, in the *elements* of its national structure that are most vulnerable to this sort of an attack. One nation is weak and vulnerable in one respect and strong in another—while the exact opposite may be true of its neighbor.

Now we have tried to show that mere indiscriminate attack of a nation is definitely not the answer. There is certainly considerable doubt that any virile nation can be defeated by any such unplanned, indiscriminate air offensive, if it is engaged in a war in which its people have a vital interest. We conclude, then, that it is essential to analyze our particular prospective enemy in each case—arrive at a true and exact estimate of its vulnerability, and then *concentrate* our attack on those vulnerable elements whose destruction will have the greatest cumulative effect in two respects. *First, on the morale of the civil populace,* by applying pressure to them through the dislocation of their mode of living and by making life under war conditions more intolerable to them than the acceptance of our terms of peace. Second, but not less important, by destroying their capacity to make war, which, of course, not only aids in achieving the first result we desire, but protects our own nation, and may well be an essential element in a sound foundation for our victorious peace.

Now the sketchy analysis that we made yesterday of our own economic structure would not apply to any other country. However, it did perhaps indicate a method to pursue in making such an analysis—the factors that we should consider. We should remember, however, that we noted that the United States and Russia are the only two industrialized powers that are built on continental proportions. In most countries the areas are much smaller, and the great city of the country is of much greater importance than is the case with us. Consider the importance to their countries of London, Paris, Berlin, and Tokyo.

London, for example, is the chief port of the British Empire. Furthermore, it is the largest manufacturing city in the empire. Many of the key industries of the empire are located within its metropolitan area, and facilities for the production of their essential products do not exist elsewhere. Over a million persons are engaged in manufacturing pursuits in London alone, out of its population of some eight million. In addition, London is the greatest commercial and financial center of the empire, besides being the directing brain for all of its scattered territories and possessions. Without a detailed analysis we cannot arrive at a definite conclusion as to just what the loss of London would mean to Great Britain, but we do know that it would be very serious indeed. Many British military writers have assumed that it would be decisive. The same is true to a great extent of Paris, which is the heart of France.

It might well be, then, that the great city would turn out to be the most important and vulnerable element in the country of our prospective enemy, when we come to make our analysis; or it might be, that pressure applied to the great city would be the final and deciding factor found to be necessary, in addition to the attack of some other vital elements. In any event so much has been written and said about the attack of cities which the School believes to be quite erroneous that it seems well worthwhile to clarify our ideas on this subject.

Therefore, this morning we will attempt to analyze the elements of a great city. We will try and select the vital points, get some idea of their vulnerability to air attack, and estimate the effects that might be expected from a properly planned and successfully executed air offensive. For the same reasons that yesterday we selected our own economic structure as an example, let us turn today to a consideration of New York City, as a typical great city. It is not intended to make out a case for the attack of New York City as a proper objective for air warfare against the United States. Our analysis yesterday indicated that other lines of attack would almost certainly be more profitable and conclusive. Since this might not be true in the case of other countries, what we are really concerned with this morning is an analysis of a great city by itself—what makes it great—what is necessary for its continued existence. If we can find this out, we should be able to apply our air offensive pressure most effectively, by directing our air attacks at the most vulnerable part of the city structure in a planned campaign whose results should be quick and decisive because each blow would augment and reinforce all the other blows to produce an accumulative effect.

Now taking New York City as a typical great city, let us for this once get off Broadway, neglect the theaters and night clubs and see what the rest of the metropolitan district really is like.

As a port New York is the largest in the world, handling 40% of our entire foreign trade. It possesses five hundred steamship piers, half of which are capable of accommodating ocean-going vessels, and it has more freight handling equipment than any other port in the world.

New York is the financial center of the world. Its financial machinery makes possible the major industrial undertakings of the nation which are, in the main, directed and controlled from this center. New York is the nation's market place.

New York is the largest industrial center in the world. Its 36,000 industrial establishments turn out about 10% of the factories of the entire United States—to a value of about 10 billion dollars a year.

In the metropolitan district there is a population of 11 million persons.

Brooklyn alone has 600,000 more people than Philadelphia. Manhattan contains 300,000 more people than Detroit. The Borough of the Bronx is about the same size as Los Angeles. The density of population over the entire district is around 200 per acre, and for Manhattan it is about 1,000 per acre. Why did such an enormous city develop here at the mouth of the Hudson River and how is it possible for such a number of people to be packed and crowded so closely into the area around this port? The basic answer can be given in one word—Transportation.

Since this is so obviously one of the absolutely fundamental elements in the existence of this great city, let us examine into it, trace its historical development and see what its influence really is.

A hundred years ago, New York was a city of some 200,000 people. The open fields started at 14th Street. There were no railroads, no street cars, no telephones, telegraphs, or ocean cables. The first steamship had made a crossing of the Atlantic only twelve years previously. The harbor was crowded with shipping. Steamboats were plentiful, some 86 plying the harbor of New York alone. Water transportation was coming ahead fast. Travel by water had become so firmly entrenched in the minds of the people, that few had any idea that it would ever be possible to travel by land, as fast and conveniently, as by water.

One hundred years ago, all of the food supply was grown within a fifty-mile radius of New York. The New Jersey farmer loaded his produce on a cart and hauled it down to the river bank opposite Manhattan, where he sold it to a broker, who after accumulating sufficient purchases, hired a boat to take it over to the other side.

Water transportation made the port of New York, thus the dawn of the railroad era found a thriving metropolis at the mouth of the Hudson. The first rail lines were of course very short, built for competition with the passenger and mail traffic, that had been monopolized by the stage coaches up to that time. By successive steps, the railroads established terminals on the Jersey shore directly opposite Manhattan, either by extending their lines, or by buying smaller rail lines. The New York Central established its all-rail line down the west side of the island. The next step was the inauguration of the lighterage service to relieve the ferry boats of the growing burden of freight traffic; then came the introduction of the car floatage in 1866 as a substitute for lighterage.

New York, therefore, is different from almost any other port, in that the great majority of its industries and steamship piers are on the opposite side of a mighty river from most of its rail terminals, and it is only within the last three decades that it has been possible to cross the Hudson River by any other means than a boat.

Legend has it that the first passenger line of transport was instituted in 1746, and ran from the Battery up Broadway to Houston St., using oxen for tractive power. Stage coaches were in use at the time of the revolution. Omnibus lines appeared in 1801. Horse-cars became a factor of importance in 1852.

The *Evening Post* of March 20, 1867 said: "The means of going from one part of the city to the other are so badly contrived that a considerable part of the working population spend one-sixth of their working day on the street cars and omnibuses, and the upper part of the island (Manhattan) is made almost useless to persons engaged in daily business in the city."

The solution to this problem was transportation.

Elevated railroads operated by steam appeared in 1871.

Cable car lines in 1886.

Electric surface cars in 1898.

Subways in 1904.

The electrification of the steam railroads and the Pennsylvania Tunnel in 1908.

In 1929 nearly one-half of the people that boarded railroad trains in the United States did so in the Metropolitan District. This does not include the rapid transit riders, whose number, of course, exceeds those of all the steam rail lines in the country combined.

Rapid Transit facilities provided 500 rides per capita in 1929 and about 8,100,000 people used these systems on the average daily in 1934.

It has been estimated that there are as many as 400,000 people underground at one time either being transported or waiting for transportation.

I would like you to bear in mind this picture of transportation within the metropolitan district as I shall come back to this point a little later.

In the meantime, however, let's turn our attention briefly to another element that is vital in the life of every great city—water supply—its water system. There is no necessity of developing the essential nature of an abundant supply of pure water, not only for drinking and cooking but for sanitary purposes and fire fighting. Without an adequate water supply there would have been no New York, nor could there continue to be a New York were it

Vital and Vulnerable   169

deprived of this essential element. Even if the supply were so reduced as to make it insufficient, it would cause the greatest hardship and would doubtless require at least partial evacuation of the city.

Let us take a look at the New York City water system and see what it is like and how vulnerable it is to interruption by air attack.

[Here the speaker refers to a display chart of the New York water system, now lost.]

There are three main sources, which altogether are barely equal to the demand. The Catskill source—510 million gallons daily. The Croton source—300 million gallons daily. Small streams and wells on Long Island—210 million gallons daily. The Catskill supply is by far the most important and furnishes over half the total supply. It starts at the Schoharie water shed 120 miles from New York. The Schoharie reservoir collects the waters of this water shed and they are taken from it through the Shandakten tunnel under the mountains and poured into Esopus Creek which conducts the waters to the Ashokan reservoir—here the Catskill aqueduct proper starts and runs over 80 miles into New York City.

The dams of these two reservoirs are of massive construction and do not offer a particularly good target for air attack. The aqueduct from the Ashokan reservoir runs along as a cut and cover aqueduct except where it passes under Rondout Creek—the Walkill River and the Hudson—where syphons are employed. It is interesting to note that this aqueduct passes under the Hudson River 1,200 feet below sea level. This aqueduct empties into the Kensico reservoir which is of sufficient capacity to hold a month's supply for New York City. From the Kensico reservoir a cut and cover aqueduct leads to the Hill View reservoir, which is merely an equalizing reservoir containing one day's supply for New York City. Here there is a connection with the Croton supply system. At the Hill View reservoir the city tunnel starts and runs for eighteen miles south, 200 to 700 feet below Manhattan and the Bronx, to its furtherest supply point, Silver Lake reservoir, on Staten Island, which is reached through a syphon under the Bay. The whole aqueduct is 92 miles long, with 35 miles of tunnel, six miles of syphons and 55 miles of cut and cover. The aqueduct in the cut and cover sections is made of concrete one foot thick at the top and four feet thick at the base of the side walls and covered with three feet of earth. It makes a mound easily visible from the air and can be located without difficulty. The aqueduct itself is about 28 feet wide and a railroad train could run through it with ease.

Now the Croton supply has two aqueducts which connect with the Catskill supply at the Hill View reservoir. The new aqueduct has a capacity of 300 million gallons daily and the old one about 35 million. They connect with the supply reservoirs of Manhattan but do not cross to Brooklyn, Queens and Staten Island, except through the Hill View reservoir connection with the Catskill system. On account of the lower level of the Croton system all water must be pumped except that of the lowest levels of the city. The Catskill system delivers water to all its distribution reservoirs by gravity. All of the Long Island subsidiary supply must, of course, be pumped.

As has been indicated, the aqueducts themselves are quite vulnerable to bombs. The Catskill aqueduct would naturally be cut between the Kensico reservoir and the Hill View reservoir, which contains but one day's supply. This operation should not be difficult with a 28-foot-wide aqueduct of practically unlimited length. While the Croton aqueducts are smaller they are equally vulnerable to small bombs and it should be quite possible to cut them.

[The accompanying chart showing the effect of cutting aqueducts is also missing.]

What would happen if these aqueducts were cut? We find that there would still be enough water available for drinking purposes but that it would be quite impossible to distribute it. Thus, while Brooklyn would be fairly well off, Manhattan, the Bronx, and Staten Island would be in a very bad condition. Mr. Nelson, engineer of the Department of Water Supply of New York City, estimated that it would be impossible to evacuate the city of New York under these circumstances, in time to prevent a considerable death toll due to thirst, on account of the complete lack of adequate distribution facilities from the small Brooklyn supply. The influence on sanitation and public health would be of the most serious nature and, in addition, the fire hazard in certain sections would be very great. While the two high pressure fire systems have an auxiliary salt water supply, these cover only the congested areas of lower Manhattan and Brooklyn. Sixty-six fires a day ordinarily occur in New York, which are controlled with only an average loss of $707.00—due entirely to the adequate water supply. Denied an adequate supply of water for fire fighting, the fire risk would be very great and considerable portions of the city would probably be destroyed.

It has been estimated that repair to the aqueducts could be effected in about one week, unless the damage was extensive. Therefore, in order to assure evacuation of the area, it would be necessary to cut the aqueducts again or to prevent repairs through chemicalization. It is apparent that if operations could be sustained, this method of attack on the city could force its almost complete evacuation with all that that would imply—not only to the city itself but to the nation. Even cutting them once would certainly deal the area a serious blow.

There is another very vital element of any great city and that is, its supply of foodstuffs. A century ago all of the perishable foodstuffs that were consumed in New York City were grown within a fifty-mile radius and were largely transported to the markets by the farmers who raised them. Let us see how far we have come from that condition with the continued growth of this metropolitan center.

The daily food of one-tenth of the people of the United States now depends upon the efficiency of the New York food distributing system. This metropolitan area alone consumes yearly about 8½ to 9 billion pounds of all foodstuffs. Let us see what this means in terms of average daily receipts of perishable foods in 1934, shown in table 5.3.

Table 5.3.		
Carloads	Item	Miles of train
800	Fruits and vegetables	7½
400	Milk and cream	3¼
132	Butter, fish, eggs, and dressed poultry	1¼
105	Dressed meats	1
215	Livestock	2
34	Live poultry	⅓
67	Frozen and fresh fish	⅔

This is equal to a train of 1,750 carloads, which at 100 cars to a train for one mile is equal to 17½ miles of train per day.

The percentage of perishables moved by motor truck is 21.8%, and with the exception of a small amount of dairy products and some 20% of the fruits and vegetables, all the remainder is carried by rail transportation. New York

City is peculiar in this respect, as it is more dependent upon rail transportation for its food than any other large city in the United States. This is due to the longer haul required from the point of origin, and the motor truck cannot be used to replace rail transportation. The foodstuffs raised within one hundred miles of New York City, each year constitutes a smaller and smaller source of supply.

Two developments have made possible supplying such a congested area with the enormous quantity of food required. These are the refrigerator car and the cold storage warehouse. In New York City there are twenty-eight large cold storage warehouses with an aggregate cooler and freezer space of 40 million cubic feet. The supplies maintained in these cold storage warehouses vary, for different items, with the seasons of the year and, for the items checked, range from one day's supply as a minimum of one item to 47 days' supply as a maximum for another—which happened to be eggs.

It is apparent that this whole system of feeding this great metropolitan area depends upon the continuance of uninterrupted rail communications into the area. With any interruption, shortages in various items would become apparent almost immediately. If the interruption could be maintained for a period of even a week or so, the area would become untenable and the population would have to be evacuated, if any way could be found to accomplish this in the absence of the continued functioning of the rail system.

[The lecturer refers to a chart of New York's rail system.]

Let us glance briefly at the rail system of New York and see what it looks like. We find that only one all-rail freight line serves Manhattan. The Long Island Railroad is confined to Long Island. The New York, New Haven and Hartford has a terminal in the Bronx and connects with the Long Island Railroad via Hell Gate Bridge. Altogether there are twelve railroads entering the district, but note that nine out of the twelve terminate their rail lines west of the Hudson River. If you will remember the length of that daily train of food—some 17½ miles—you will at once appreciate the vital importance of these terminals to the food situation of New York City. Any cut that will deny the use of these terminals will practically eliminate the railroads as a factor in the movement of freight into the metropolitan area. The terminals are essential to operate the car float and pier system by which freight cars are brought across the river on the floats and unloaded directly onto the piers. If rail lines were cut so as to deny the use of the terminals, it would be necessary to unload the cars at sidings on the other side of the break and transport by truck into the city itself. Aside from the traffic congestion and general difficulty of doing this, I think we may conclude that the problem could hardly be solved in this manner, since outside of the terminals there would not be sufficient suitable sidings at which the 17½ miles of cars could be unloaded. It is not too much to say that the entire marketing, transportation and food storage systems in New York City depend upon the continued and uninterrupted use of the rail terminals.

We haven't the time to go into an analysis of what could be required in the way of air attack to cut the railroads off from the use of these terminals. However, just a glance at this small map we have here shows some nicely concentrated vulnerable objectives where the railroads pass over the wide tidal rivers on bridges. I think it is fairly safe to assume that this task would probably lie within the capacity of quite a reasonably sized striking force. If an adequate analysis should confirm this conclusion, it is apparent that New York City could quite easily be eliminated by an air offensive directed against this particular element.

Now, no doubt, careful study and analysis of the great city of New York would indicate many other vital elements that might be worthy of our attention. However, our time is growing short, and we will, therefore, consider as a final vital point, an element which our prior discussion indicates may well be decisive—electric power.

Greater New York City has two distinct power systems, between which there is no inter-connection whatever. One lies on the Jersey side of the Hudson and the other on the New York side. Now the Jersey system, including Staten Island, has a total of eleven steam generating plants. However, the bulk of the generating capacity is concentrated in the three plants of the New Jersey Public Service Corporation on the salt meadows near Newark. These three plants develop over 75% of all power generated on the New Jersey side of the river, where most of the heavy manufacturing industry of the New York district is located. An inter-connection is maintained between these plants and the Pennsylvania Railroad for stand-by services only.

On the New York side of the river there are a total of twelve important steam generating plants; in fact, in the whole metropolitan area, within a radius of 22½ miles from the city hall, there are only twenty-six generating plants altogether.

The situation in New York City is considerably complicated—due to the presence of two systems, one using 25 and the other 60 cycle current. Some of the plants furnish one type only and some both types, but the systems cannot be inter-connected due to the differences in equipment.

In the whole area there are eight steam generating plants which furnish power for the traction systems: Interborough Rapid Transit—Brooklyn-Manhattan Transit Corporation—The 8th Avenue Subway—Hudson-Manhattan Lines—Long Island Railroad—Pennsylvania Railroad—and the electrified portions of the New York Central and New York, New Haven and Hartford Railroad.

But perhaps this limited number of generating plants is offset by outside sources of power? Upon examination we find that there are only two main transmission lines running into the area—one in Jersey and one in New York. The capacity of these transmission lines is strictly limited and the total power that can be moved in is equal to less than half the power that can be generated in one plant alone of the Brooklyn-Edison Corporation.

Furthermore, we discover a very interesting situation. These outside power sources are tied directly in with the generating plants. The outside high voltage is brought to the steam generating plants for transforming and distribution, due to the saving in duplication of facilities which this system makes possible. We find, therefore, that the generating stations themselves are essential for the use of this outside power.

Now what about the vulnerability of electric power in the metropolitan area? Certainly, I think, we are all prepared to admit that a direct hit from one bomb of the proper size will destroy one of these steam generating plants. We have seen from our previous investigation that from 9 to 18 months would be required for its replacement.

The figure entitled "The Aerial Bomb vs Public Service Electric Power in the New York City Area" shows a graphic picture of what such hits, by a limited number of bombs, would mean to the industrial and public service power of the metropolitan area. With this limited number of bombs, accurately placed, less than 10% of the public service power is left available and this is at Staten Island, Amboy, and Raritan.

The figure entitled "The Aerial Bomb vs Traction Electric Power in the New York City Area" shows the effect of a strictly limited number of bombs accurately placed on the sources of traction power of the metropolitan area. The only power remaining is a small amount in one plant for the New York, New Haven and Hartford Railroad, and one small plant for the Brooklyn-Manhattan Transit Corporation. We see then that seventeen bombs, if

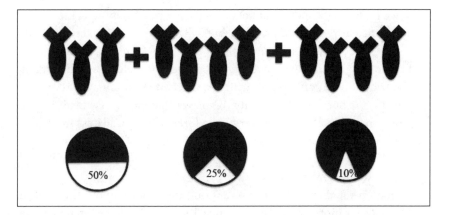

"The Aerial Bomb vs. Public Service Electric Power in the New York City Area" illustrates the impact of a limited number of accurately placed bombs and the resultant reduction in the overall available public service electric power for the New York City metropolitan area: three bombs reducing power by 50 percent, four more bombs reducing power to 25 percent, and four more bombs reducing power to 10 percent.

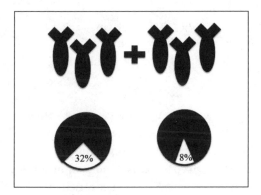

"The Aerial Bomb vs. Traction Electric Power in the New York City Area" illustrates the impact of a limited number of accurately placed bombs and the resultant reduction in traction power for the New York City metropolitan area electric rail system: three bombs reducing power to 32 percent and three more bombs reducing power to 8 percent.

dropped on the right spots, will not only take out practically all of the electric power of the entire metropolitan area but will prevent the distribution of outside power! Let us add one bomb for the Brooklyn-Manhattan Transit plant and make it eighteen. Two squadrons of bombers, therefore, would certainly do the trick if they could bomb with 100% accuracy. You may add on your own factor for bombing probability. *Whatever* that factor may be, it is quite apparent that it would take no very large force to practically assure depriving this whole great metropolitan area of all sources of electric power for a period of many months.

What would this mean?

First, no rapid transit; the subways out; the tunnels under the Hudson and East Rivers inoperative. Pennsylvania Terminal dead; the Grand Central Terminal partially out; the elevated railroads stopped; no trolley cars.

And what about the influence on vehicular traffic? The Holland tunnel requires electricity for ventilation. Without ventilation it cannot be used.

There would be no vertical traffic. More passengers are carried in elevators than by horizontal means. This would deprive the great buildings of New York of their usefulness.

What of the influence on ports?

The unloading of vessels depends on cranes and other apparatus operated by electricity. The reduction in the usefulness of the port of New York is obvious.

What of food supply?

All the perishables of New York are stored in cold storage warehouses operated by electricity. We might expect wholesale spoilage of food and inability to maintain an adequate supply on hand. All the bread of New York City is made in modern electric ovens. One load in every six baked in the United States would not be baked without electricity in New York.

What of the water supply?

You will recall that both the Croton water supply and the Long Island supply have to be distributed by pumping. This pumping is of course done by electricity. At one stroke then the distribution of one-half the water supply of New York City is rendered impossible with all that this entails.

And what of the fire hazard?

The high pressure fire systems, upon which reliance is placed in the congested areas, would be inoperative without electric power.

And what about the manufacturing industry?

270,000 electric motors, representing nearly 65% of all the horse-power used in manufacturing in the New York industrial area, would stop.

There would be no street lights, no electric lights in the home, no cooking on electric ranges, no household appliances. People would not know what had happened. There would be no newspapers. The congestion of vehicular traffic would make it impossible to move. Even if the news were broadcast, people would be unable to receive it for radios would be dead.

At one stroke, the industry, the home, the entire machine that we know as New York City, could not function. Certainly the bulk of it would have to be evacuated. At least until numbers were reduced back to those of the horse car and ox team era of the old days.

Remember that New York was built because of transportation. Modern transportation facilities alone make its continuation possible. If these are taken away it will no longer be possible for a city like New York to exist.

I think our analysis has been carried far enough to illustrate how we should go about planning the air attack of a great city, should we ever be called upon to undertake such a job. Incidentally, I think, it has resulted in a rather graphic picture of the vulnerability of a typical great city to planned air attack.

Let us sincerely hope that in the event a hostile air force ever comes within striking distance of New York City, that it will be directed by one of those outstanding military thinkers who is perfectly sure that air offensives against great cities can never be effective.

Are there any questions or comments? Thank you.

Comments on "National Economic Structure" and "New York Industrial Area"

"National Economic Structure" distinguished American thinking on strategic bombing from that of the Europeans by deeming Douhet's emphasis on bombing civilians not only immoral but, moreover, ineffective. Air operations in Spain and China indicated that the resolve of the population, once stiffened, was not easily swayed. By contrast, ACTS emphasized attacking a modern industrial nation's infrastructure to reduce the enemy's capacity to wage war while simultaneously applying economic pressure on the population. ACTS faculty, like other interwar military theorists, were heavily influenced by how World War I ended. Douhet believed that the direct bombing of cities would cause the collapse of the will of the enemy population, as had happened in Germany. Fairchild, however, argued that the collapse of the enemy's will would be best achieved by bombing an enemy's wartime economy. The cumulative effect of aerial bombardment would more efficiently reduce the enemy's material means to fight than that of conventional land battle, while simultaneously subjecting the population to severe economic deprivation. It was this combination of lacking the material means to continue the war and the suffering endured by civilians that would compel the enemy to capitulate.

The specific targeting of an industrial nation's economy entailed two steps: first, identifying what key factors of production were vital and second, determining the location of concentrated chokepoints vulnerable to air attack. Fairchild believed the information needed to target fixed infrastructure was readily available, although World War II and the Cold War would later prove this optimism to be misplaced. Overall, however, Fairchild succeeded in identifying three critical requirements for any modern industrial state: raw materials, transportation, and power.

With a grand strategy of isolationism in the United States, the ACTS faculty was prevented from obtaining intelligence on potential adversaries. In the second hour of the "National Economic Structure" lecture, Fairchild instead analyzed how the United States might itself be vulnerable to air attack. Given the US abundance in most raw materials and the multiple sources for those materials that were in short supply, Fairchild did not view it as practical to try to disrupt imports to the United States. He did, however, point out that import-dependent countries such as Great Britain, Germany, Italy, and Japan might be vulnerable to interdiction or blockade. Where Fairchild believed the United States to be most vulnerable was in its transportation networks, its concentrated petroleum refineries, and its electrical power stations. All three of these factors of production would later be identified by the United States Strategic Bombing Survey after World War II as key vulnerabilities in the Nazi German economy.[7]

Fairchild, in "New York Industrial Area," provided a blueprint for how to identify the vital and vulnerable economic nodes of a great city. Though his assessment was not intended to be comprehensive, he examined three systems in some detail: water supply, transportation, and electric power. With the exception of a comment on the potential need to reattack aqueducts, Fairchild largely ignored the question as to the likely resilience of an enemy's wartime economy and the impact this might have on the operational requirements to continually revisit targets.

Together, these two lectures captured the core, foundational idea of the ACTS theory of strategic bombing: that the high-altitude daylight precision bombing of vital and vulnerable nodes of a modern nation would produce economic paralysis. Analysis after World War II would show that Fairchild was largely correct to emphasize attacks on transportation and power networks. Where he was wrong proved to be in errors of omission rather than commission. He ignored the potential resilience of a nation's wartime industrial production and did not consider how economic paralysis would lead to the desired political outcome. This missing link in the ACTS theory of victory would be revealed later in World War II when the collapse of the German and Japanese economies failed to prompt the surrender of either the Nazi regime or the Japanese Empire.

Despite this major shortcoming, the Air Corps had gone far beyond merely theorizing about strategic bombing to producing the B-17, mounted with the Norden bombsight, as a means to conduct precision strikes and developing a method for determining the bombing requirements for an air campaign against Germany and Japan. In the next chapter, Fairchild concludes the Air Force course by considering the critical decision as to whether to attack the enemy's military or its economy.

6

What to Target
The Economy or Military Forces?

PRIMARY STRATEGIC OBJECTIVES OF AIR FORCES

In the final lecture of the Air Force course, "Primary Strategic Objectives of Air Forces," Muir Fairchild concludes that the fundamental decision for air warfare is whether to attack an enemy's national economic structure or its military forces. He observes that attacking the economic structure is a purely offensive action, while attacking enemy forces has both offensive and defensive objectives. Which objective to pursue depends as much on the vulnerability of one's own nation as it does on that of the opponent. The nation least vulnerable to attack should select the enemy's national structure as its primary target. An additional advantage in attacking the national economic structure is that intelligence on fixed, nonmilitary targets is available in advance during peacetime, unlike the intelligence required to identify mobile military forces under wartime conditions, making target selection much easier.

Alternatively, for continental nations whose primary concern is the threat of invasion, the proper objective should be an attack on the enemy's military forces. Fairchild refers back to the threat to France from the German breakout along the Western Front in early 1918. It would have done little good to be bombing factories while German soldiers marched through Paris. Likewise, for a revisionist state such as Germany, which counted on victory through conquest, its air force would also be best employed against the forces of its enemy. Fairchild presciently argues that, for cases in which an air force cannot reach an enemy's national structure, as with the United States against Japan, a direct attack on enemy forces would be required to advance air basing to positions from which bombers could then reach the enemy's homeland.

Fairchild concludes his lecture by restating a final advantage to attacking national economic structure. Technologically sophisticated weapons were playing an increasingly important role in modern war. Paralyzing a nation's wartime industrial production would prevent it from arming its military forces. Unless a country has sufficient forces in being at the outset of war, an attack on its national economic structure would deny the enemy both the capability to defend itself and the means to sustain its population. Such a situation would reduce the enemy's hope of a military victory while simultaneously imposing costs on its population for continued resistance.

Primary Strategic Objectives of Air Forces

Major Muir S. Fairchild

April 11, 1939

Gentlemen, in this course on Air Warfare we have been considering the background and underlying principles upon which correct decisions regarding the proper employment of air forces must be based.[1] It may seem strange that we should find it necessary in this course to devote so much time to the study of objectives, while the courses in the ground arms are able to avoid the subject almost entirely. However, we must remember that the objectives of ground forces have been well established by long experience with these forces. They have a rather restricted field of choice in any event. As ground forces must be forever concerned with the holding and occupation of territory, their objectives must continue to be largely topographical.

When we visualize the employment of air forces, however, we immediately find before us a heterogeneous collection of minor objectives—airdromes and port facilities—troop columns and battleships—rail centers and artillery batteries—factories, bivouacs, bridges—cities, and as some would have it, even the infantry line in battle! The attack of any one of these multitudinous, crowding objectives may be expected to have some virtue, however slight. Also we may expect each particular objective to have its insistent advocates, if the present variety of thought on the question continues. Certainly there is as yet no unified opinion among military personnel in general on this important fundamental question.

One great reason for this condition is the attempt to fit this military factor of air power—air warfare—into the old picture of traditional warfare, without regard to the proper application of the real fundamentals that have held true throughout the history of war. Now certainly we have not presented any new or revolutionary principles; but we have tried to show in a broad and general way what the School believes is the proper application of those fundamentals to the conduct of air warfare.

Let us briefly summarize the basic points that have been brought out before proceeding to consider the question of the primary objectives of air forces.

In the first place, the ultimate aim of all military action in war is the breaking of the hostile will to resist. Surface forces, due to the inherent characteristics of surface forces, are obliged to accept an intermediate objective—the armed force objective. They are obliged to defeat the corresponding types of hostile armed forces before they can proceed on to their ultimate objective. Otherwise they are normally open to defeat by those hostile forces, through action directed at their vulnerable lines of communications. Now air forces are fully capable of acting against the armed force objective—incidentally, not only the corresponding *type* of armed force, but against *all* types of hostile armed forces. But beyond and above this capability, if they possess proper strategic air bases, air forces may act directly, immediately and continuously against their ultimate objective—the direct attack of the structure of the enemy nation.

Here, then, we come to a fundamental difference between surface forces and air forces. The primary strategic objective of a land force is the hostile land force—to assure the defeat of the opposing land force. The primary strategic objective of a naval force is the defeat or the containing of the hostile naval force. *However,* the primary strategic objective of an air force is not necessarily the hostile air force. True, it *may* be the hostile air force, but it may also be the hostile ground force, the hostile naval force; or it may perhaps ignore all such intermediate objectives and make directly for its ultimate objective, the direct attack of the national structure of the enemy nation.

It is obvious that this complete choice does not *always* exist, as was demonstrated, for example, in the case of our being on the strategic defensive without allies. The reasons, of course, are to be found in the present range of aircraft and the lack of properly placed strategic air bases. We should not lay too much stress on this case however. It certainly is not the condition as between two such nations as France and Germany, for example, where there is complete freedom of choice. In their case *all* the elements, not only of all the armed forces but of the respective nations themselves, lie well within their radius of operations from existing air bases. It very likely will not actually be the condition in our own case, having regard for the world situation today. The interdependence between nations and the alignments which are everywhere apparent, would seem to make it unlikely that we will be involved in any major war without the participation of other nations, not only on the enemy side, but as allies in our own effort.

If allies are present, they may be able to furnish us with the strategic bases that we require. Hence we should not definitely conclude that this freedom of choice, as regards the primary strategic objective of our air forces, is not liable to be open to us. It seems quite possible, therefore, that if we are involved in a major war, within the predictable future, we may be faced with some choice between the armed force objective and the enemy national structure objective.

This choice between these two great categories, as the primary air force objective, is the most fundamental decision involved in air warfare and it will doubtless be correspondingly difficult to make. Were there no factors to consider, other than the effectiveness of an air offensive in achieving its final purpose—the collapse of the enemy will to resist—there would be little difficulty in making the choice. Obviously and certainly we would choose to attack the enemy national structure as our primary objective.

However, there are, of course, other factors to consider. While we plan destruction for the enemy, we must not forget that he, in all probability, is planning very much the same sort of fate for us. This decision, then, whether we will choose the enemy national structure or the attack of the hostile armed forces as our primary strategic objective, is a matter of very considerable moment. If we decide to attack the enemy national structure directly, neglecting the enemy armed forces as an objective, we may open up our own nation to the same sort of an offensive. You will certainly have become convinced by your studies at the school, if you were not already certain of the point, that the only method known at present by which we may obtain relative security from air attack for our own nation, as well as for our armed forces, lies in directing an air offensive against the hostile air force. Thus we may under some circumstances be forced into a decision accepting, as an intermediate objective, the attack of the enemy air force, or perhaps his other armed forces.

It would certainly be fine if there were some simple rule that could be quoted which would cover the matter of how this decision is to be made. Life would certainly be much simpler if the conduct of war could be formulated under a set of rules. But unfortunately that is not true, and probably never will be true. In this case as in most of the other major decisions of war—it depends on the situation. In fact, the whole art of air strategy consists in properly evaluating all the factors of the situation and deciding against what hostile objectives our air offensive should be directed to so exploit the peculiar powers of the air force as to secure its maximum contribution to the national aim.

Since it is impossible to establish a rule applicable to all situations, let us examine some of the important factors that must be taken into account in arriving at this decision.

The point in question is whether, in the final analysis, the more certain path to victory, even though somewhat longer and more difficult, does not lead through the defeat of the hostile armed forces. The *end* sought in each course of action is the same. In one case it is intended to remove the forces that may successfully oppose our offensive action. We would then achieve a position that would permit the unhampered application of pressure upon the hostile nation. The other theory prescribes the application of that pressure in the first instance, and the neglect of the hostile armed forces as an objective of other than casual concern.

The direct and exclusive attack of the structure of the enemy nation is of course purely offensive action. It must succeed quickly, as it permits the air force of the enemy to operate with the utmost freedom. It may give him the opportunity to strike, in his turn, similar targets within our own nation as well as to operate against our air force with the idea of defeating it directly. On the other hand, the selection of the hostile air force as the primary objective has a dual purpose. It seeks an ultimate victory through the defeat of the enemy's air force. In the meantime it provides the maximum possible degree of security of all objectives that the opposing air force would seek to attack, including not only objectives within the nation but also objectives vital to the continued action of our own armed forces.

Now it seems from this discussion that the governing factor must be the relative degree of vulnerability of the opposing nations and of their various classes of objectives. Certainly this degree of vulnerability will never be exactly the same and usually it will be very different.

Let us take the simple case where opposing nations each select the enemy national structure as the primary objective of their respective air forces. If each air force continues to pound away at the hostile national structure, without regard to the hostile air force, and each of the air forces is able to conduct operations with the same degree of efficiency, it is obvious that the nation which is most vulnerable to air attack will be the first to collapse in such a duel. Of course this situation is too simple, for it is unreasonable to suppose that the air forces of any two nations will be of precisely the same strength and efficiency, or will be placed with equal advantage for the conduct of air operations. However, here we seem to have the underlying factor in making this decision. The nation which is the least vulnerable to the attacks of its enemy,—the nation which feels that it is in a position to carry an air offensive to the enemy nation with prospects of early and decisive success before the enemy can effectively counter the attack,—will select the enemy national structure as its primary objective. On the other hand, the nation that feels itself to be at a disadvantage in such a duel will in all probability select the hostile air force as its primary objective.

We may expect that, under normal circumstances, where it is possible to do so, the *initial* attacks of *any* aggressor nation will be delivered against the hostile national structure. This decision will be due to the resulting shock effect, the vital importance of certain installations to a nation in arms, the degree of facility with which these installations may be destroyed, and to avoid the preliminary air reconnaissance which would otherwise by necessary and which would disclose the intentions of the aggressor nation. Complete information of the vital nonmilitary installations within a nation will usually be available during peace, while the detailed information necessary for counter-air force action will normally have to be obtained by reconnaissance after the outbreak of war. Of course, we may expect that operations would be directed against the enemy air force at any time during the progress of the war should it present a concentrated, vulnerable objective which could be destroyed or seriously damaged in one, or a few, quick operations. Furthermore, the *superior* air force might well commit to counter-air force actions only that part of the force required to obtain the necessary degree of neutralization, while the remainder, or "marginal force," is employed in a direct offensive against the enemy nation itself. While such action might seem to be a violation of the Principle of the Objective, we must remember that we are also concerned with the Principle of Security. Such a decision, therefore, may be fully in accordance with the Principles of War, however much it may appear at first glance to be a violation of them.

So far we have not discussed the possibility that hostile surface forces might constitute our primary objective. Now it must be recognized that when a nation is on the strategic defensive, a hostile surface force might, at any particular time, offer the greatest threat to its security. This, of course, would be particularly true in the case of nations with contiguous borders. We can easily see that in such a case, if our own surface forces should be overwhelmed, a powerful offensive by a rapidly moving land force might constitute a decisive threat. It might well threaten to defeat our nation quickly, and before any action that the friendly air force could take against the enemy nation, would become effective. In such a case there is no doubt that the hostile land force becomes the primary strategic objective of the friendly air force, since this employment would constitute its maximum contribution to the national aim at that time.

As an example of such a situation, consider the German break-through on the Western Front in March of 1918. This break-through was practically complete. For a time the road to Paris lay pretty well open before the German troops. Had Paris fallen and the British forces been split off from the French it is very probable that it would have meant the end of the war for France. The situation was desperate for the Allies. Everything that could be constituted as a reserve was thrown in to close the gap—incidentally, including all the aircraft that could be brought to bear. Under such circumstances, had there been an independent air force operating against Germany directly, and expecting to achieve victory by its operations within a short time, should it not also have been thrown into the gap? I think the answer is obvious. Certainly it should have been so employed, for here was a decisive threat that might well mean complete disaster. By employing it against that decisive threat until the threat was removed or countered, it would undoubtedly have been making its maximum contribution toward the Allied aim in the war at that time.

There is the reverse of this case, also, in which primary reliance for victory must be placed on the friendly land force, and where the maximum contribution of the friendly air force can best be made by directly assisting the operations of the friendly land force. In such a case we find that the defeat of the enemy land force might well become the primary objective of the friendly air force.

We find an excellent example of such a situation in the operations incident to the Italian conquest of Ethiopia. Reliance for the seizing and pacification of Ethiopian territory, preliminary to its incorporation into the Italian Empire, obviously had to be placed upon the Italian ground forces. Furthermore, there was no integrated national structure or Ethiopian air force against which the Italian air force could have been employed. Under such circumstances we conclude that the employment of the Italian air force directly to assist the Italian ground forces, to defeat the Ethiopian ground forces, was the proper and logical employment since this constituted its maximum possible contribution to the Italian war aim.

When a nation is on the strategic defensive, it might be that its air force would have no objectives within range from its available air bases, other than those hostile naval forces that placed themselves within effective radius of action. In such a case, obviously the hostile naval force becomes by necessity the primary strategic objective. Such a case is illustrated by our own situation, if we should be on the strategic defensive without allies, and hostile air forces should not be present.

A situation might arise where initial reliance would have to be placed on our naval forces, in the strategic offensive, to conduct operations leading to the securing of bases for the future operations of our land or air forces. In such a situation the initial objective of our air forces would be the same as that of our friendly naval forces—the defeat or containing of the hostile naval forces. Of course, should those hostile naval forces be supported by hostile air forces, the maximum contribution of our air forces would be made by employment against the hostile air force as an intermediate objective, since such action would be beyond the powers of our naval surface forces.

An illustration of such a situation is to be found in the step by step progression across the Pacific Ocean toward Japan that has sometimes been proposed, in the event of war with that power without allies on our side. In such operations it is proposed that we advance outward from Hawaii, seizing island after island and utilizing each in turn as a base from which to conduct operations against the next one in order. In this case it is apparent that the objective of our air forces is the same as that of our naval forces—the defeat of the hostile naval forces to permit the step by step advance to continue. It is equally apparent that, if hostile air forces are present and constitute a serious threat, our air forces provide the only means available to oppose that threat and that their maximum contribution would lie in removing the threat before proceeding against the hostile naval forces.

The decisions seem quite obvious in these situations, where the hostile surface forces are selected as the primary objectives of our air forces. In fact they are, in each case, forced upon us by the situation. We have little or no choice. The situation itself dictates clearly and unescapably the primary objective of the air forces.

However, when we come to select the enemy national structure, or the hostile air force, as our primary objective, when both these classes of objectives are open to us, the choice is not so clearly indicated and we must consider a great many factors. In fact, it seems apparent that the only basis upon which this decision can logically be made, is upon a thorough national estimate of the situation. In such an estimate we must consider in great detail the relative vulnerability to air attack of the nations involved, and the relative ability of the two nations to conduct the required air operations. Furthermore, the relative vulnerability of the two air forces must be carefully weighted, since this factor, also, will seldom be the same for both belligerents.

The process of making such estimates of the relative vulnerability of nations has already been indicated. Obviously it is a task requiring some time and a properly trained and balanced staff, if a satisfactory job is to be done. There seems to be no short cut unless the situation itself is clear and simple.

There are many factors that must be evaluated. This course has attempted to point out what those factors are and to indicate how they may be expected to influence the decision.

We will try and sum up the general effect of those factors and arrive at as concise a statement as possible of a doctrine for the employment of air forces.

War, throughout the ages, has become more and more a question of machines. It is not to belittle the human element, or to disparage the qualities of courage or leadership, to recognize that in war today machinery is paramount. War without machines is impossible. The significance of this development, in its bearing upon the employment of air forces, has not been sufficiently appreciated.

The armed forces all require machines of war, whether the vast intricate battleship machine or the simple basic infantry rifle machine. Each of these war machines is the product of other and prior machines. Machines make machines, and the machines which *make* are even more elaborate and complicated than the machines which are made. A multitude of patterns, jigs, gauges, forgings, castings, and stampings,—a long process of turning, milling, cutting, boring, drilling, and machining, of work in the forges, presses and lathes, go to the making of any modern arm of war. Even the infantry rifle and the ammunition it uses are the result of some thousands of mechanical operations.

Now the machines which are made and in the hands of the armed forces are mobile; they are hard to locate; and they are dispersed. The machines which make these machines are fixed; the location of their vital points and motivating powers are known; above all, they are concentrated and they cannot be dispersed. Here, then, we can deduce at least a minor principle: We cannot only destroy the machines, which alone make war possible, more effectively, if we strike before they are brought into action by the armed forces. We can, by striking correctly, prevent them from ever coming into being. By selecting proper objectives we can destroy the machines that make the machines and so destroy whole generations of war machines yet to come.

By so attacking our enemy, we strike at the heart of his war strength. We strike at his power to create, maintain, and replace his armed forces.

Here, then, we have the armed force objective on a vastly expanded scale. More vulnerable because it is more concentrated, without mobility, without inherent power of defense, and not designed or constituted to absorb the blows of the attacker. And who would say that even the defeat of an enemy's armed force in battle would create a morale effect comparable to the destruction of those war industries which alone make the creation, the maintenance, and the replacement of those armed forces possible? A nation may still find heart and courage to resist after a military disaster. No hope remains for the nation deprived of its power to maintain, to reinforce, or to replace its armed forces.

It has already been shown that the destruction of the capacity for war of a nation can be combined with the direct application of pressure to the civil population without loss of efficiency to either purpose. The industrial mechanisms which provide the means of war to the armed forces and those that provide the means of sustaining a normal life to the civil population, are not separate, disconnected entities. They are joined at many vital points. If not electric power, then the destruction of some other common element will render them both inoperative at a single blow. The nation-wide reaction to the stunning discovery that the sources of the country's power to resist and to sustain itself are being relentlessly destroyed can hardly fail to be decisive.

The conclusion seems to be inevitable that under the Principle of the Objective, here is, in every case, the ideal primary strategic objective of the air force. It utilizes to the full the peculiar capabilities of the air force and would normally constitute its maximum possible contribution to the national war aim. Normally, only two considerations should prevent its exclusive selection. First, sheer inability of our air force to accomplish the task. Second, action found essential, under the Principle of Security, to provide the required degree of protection to our armed forces and our own national structure.

The employment of air forces against basically vital keys or links of the hostile economic structure combines the advantages of the enemy national structure objective and the armed force objective. Only those armed forces *in being* that are fully supplied and fully equipped escape the influence of such employment, if successful. This escape can only be temporary at best, for no nation in the world can afford to maintain in war reserve, the supplies, the munitions, the replacement equipment, to offset the terrific consumption which begins with war.

Hence the decision resolves itself into a relatively simple question. Can our nation and our armed forces withstand the action of the enemy's armed forces *in being*, until our air forces can exert the pressure necessary to cause the collapse and capitulation of the enemy nation? If we decide that they can do so our decision must be to exploit the peculiar powers of our air forces to the utmost by employing the air forces so as to obtain their maximum possible contribution through the direct attack of the enemy national structure. While considerations of security or sheer inability may force us to postpone this action, the first opportunity which promises success should be seized upon to initiate this employment.

Now let us see if we can clarify the results of this discussion by briefly considering an actual situation. We will take a situation which might, perhaps, be real enough one of these days.

General situation.

Due to continued developments in the Far East the United States has declared war on Japan. Russia, seizing her chance, has allied herself with us. Our War Department has ordered the available striking units of our GHQ Air Force flown to bases to be furnished by Russia, in Siberia.

As staff officers, let us consider the situation that confronts us on our arrival and prepare ourselves to make recommendations as to the proper employment of our GHQ Air Force.

We find ourselves located in prepared bases furnished by Russia, just to the north of Vladivostok, and supplied by the Trans-Siberian Railway. Large Japanese and Russian ground forces are in contact along the Amur and Ussuri rivers. The situation is static and neither side has as yet gained any advantage. The Japanese air force is located at bases within Japan proper and at their forward bases in Manchukuo. The Japanese air force is being employed to directly assist her ground forces. The location of occupied Japanese airdromes has been determined and is furnished us by the Russians. Japan is still engaged with China to the south. The Japanese ground and air forces are being supplied by ocean transport across the sea of Japan, convoyed and screened by a portion of the Japanese Navy. The bulk of the navy is screening the Pacific coast of Japan.

The range of our striking force units is such that all of these elements and all of the vital elements of the Japanese homeland are open to our attack. We have a complete freedom of choice as between the following major classes of objectives:

We can employ our striking force to directly assist the Russian ground forces in their effort to overwhelm the Japanese ground forces and drive them south out of Manchukuo. This may be done either by direct employment near the combat zone or by cutting vulnerable lines of communication far to the south. We can employ our striking force in counter-air force action, directly, to eliminate any future threat from the Japanese air force. We can employ our striking force against the transports and their convoying naval forces to prevent the supply and maintenance of any of Japan's armed forces on the mainland. We can attack the naval bases which are the source of Japan's naval strength. We can reach out with our longer range equipment to the bulk of the Japanese fleet off the coast of Japan. If this attack is successful it will reduce her naval strength to a point where our fleet would have an overwhelming advantage even in these far distant waters.

Finally, we have the choice of launching a direct offensive against the structure of the Japanese nation with the joint purpose of eliminating her capacity to make war, at the source, and at the same time of applying the most intense pressure directly against the Japanese population. The Japanese, we note, are in this situation limited in their choice of objectives to the armed force objective, either the Russian ground forces, or the combined air forces, simply because no other objectives lie within their radius of action from their available air bases.

Now what must our choice be from all these classes of objectives? From the situation as given, if our discussion this morning has any reality, it seems clear that the maximum contribution of our air force toward the attainment of our national aim in this war, lies in its employment directly against the structure of the enemy nation, ignoring all the intermediate objectives that present themselves.

Now, however, let us change this situation a little. We find that the Russians have not provided any bulk of supplies for our operations, but intend to bring them in over the Trans-Siberian railway. Word is received from the Russian High Command that the Japanese ground forces have overwhelmed the Russian forces at one point and are pushing rapidly forward to seize and cut the Trans-Siberian railway. The Russians express doubt that available reserves can prevent the accomplishment of this purpose and request that our air force be employed to directly assist them to hold back the Japanese. What is the decision to be in this case?

Or, let us take a somewhat similar change in the situation, so that the Japanese air force, having discovered our bases, is being employed effectively against our air force. Our losses are mounting. It is apparent that we cannot accomplish our purpose before the capacity of our force will be reduced too far to permit us to succeed. What is the decision here?

It seems clear from our discussion this morning that, under such circumstances, we must postpone our attack of our selected primary objective and under the Principle of Security (not under the Principle of the Objective) turn our attention to the elimination of these major threats. Once they are eliminated, or sufficiently reduced in importance, however, certainly we would once more take up the direct attack of our primary strategic objective—the National Structure of the Enemy Nation. Under such a general situation as has been pictured, such employment of our air force would surely constitute the maximum possible contribution of which it is capable, toward the achievement of our national aim in war.

Are there any questions or comments?

COMMENTS ON "PRIMARY STRATEGIC OBJECTIVES OF AIR FORCES"

In this final lecture, Fairchild continued his discussion from the previous chapter on the critical question of targeting. In "National Economic Structure" the choice lay between attacking the civilian population or the economic infrastructure. In this lecture, the decision shifted back to a more fundamental consideration of whether to target enemy military forces or economic infrastructure. The ACTS solution was to attack national economic structure, which Fairchild proclaimed to be the ultimate objective of an air force. Enemy military forces were only an intermediate objective to "the ultimate aim ... the breaking the hostile will to resist."

Enemy forces should be avoided except under certain conditions. The first caveat allowed for striking an enemy likely to quickly win the war with its forces in being. Fairchild argued that this was unlikely as nations could not afford the war reserve required for the attritional land battle needed for a modern industrial nation to invade another. A second caveat must be accorded when a nation's own economy was more vulnerable to air attack than that of its opponent. This was fortunately not likely to be the case for the United States, given its dispersed production and extreme distance from potential opponents. The third, and most likely, exception to targeting a nation's economic structure was if the air force could not reach the enemy's industrial heartland. This, Fairchild pointed out, was likely to be the case with the United States against Japan. Unless allies could be found in the region, the occupation of foreign territory would be required for US air forces to reach the Japanese homeland. The final caveat was if the enemy had no industrial capacity to target, as with the Italian invasion of Ethiopia. It would later prove tragic in both the Korean and Vietnam Wars when US airmen did not heed this advice.

The most significant criticism of this lecture has to do with errors of omission. Fairchild does not consider how war aims might affect a nation's willingness to endure the collapse of its economy and the suffering of its people. As World War II would demonstrate, a state may prefer to resist even when faced with economic collapse and imminent military defeat if the alternative is unconditional surrender and the demise of the state. Fairchild also does not consider how a powerful government, such as Nazi Germany, might compel the population to continue to fight even under severe pressure from strategic bombing.

Combined, the ACTS lectures presented in chapters 1 through 6 summarize a uniquely American theory of strategic bombing that would soon be tested in World War II. The lectures on operational requirements for conducting high-altitude daylight precision bombing offer not so much original thought from ACTS faculty as a reflection of how the Air Corps was thinking corporately about how to conduct independent air operations. The originality can be found in their connection of means to ends, that is, in articulating the causal logic of how HADPB operations would lead to desired war outcomes: the bombing of vital and vulnerable industrial nodes would paralyze an economic structure to the extent that the enemy would lose the capability and will to continue.

The rapid expansion of the Air Corps in anticipation of war brought an end to the school and, had the story of ACTS ended there in the summer of 1940, its theory might never have been tested. In August 1941, however, members of the ACTS faculty had the unique opportunity to create the US air campaign plan for World War II. The next chapter describes the contribution made by Harold George, Ken Walker, Haywood Hansell, and Laurence Kuter in making HADPB a central component in the United States' plans to defeat Germany and Japan. The chapter then assesses how well the theory performed when put into practice.

7

High-Altitude Daylight Precision Bombing in World War II

The Air Corps Tactical School lectures in the previous chapters present a uniquely American vision of strategic bombing focused on attacking the vital and vulnerable nodes of a nation's economy. High-altitude daylight precision bombing might have remained a theory only, however, had it not been for the arrival of World War II, a conflict that provided airmen the opportunity to test their hypothesis that an air force could independently and decisively win in war. In anticipation for the coming war, the Air Corps accelerated efforts to educate its airmen on strategic bombing theory. The curriculum was truncated to three abbreviated twelve-week courses taught from 1939 to 1940, during which time ACTS accommodated three hundred of the four hundred senior officers who had not previously had an opportunity to attend.[1] The Air Corps then discontinued the course, but a core group of its "Bomber Mafia" faculty would soon be translating theory into practice planning the air war for the newly expanded US Army Air Force.

In the summer of 1941, Lieutenant Colonel Harold George received the assignment as chief of the USAAF's new Air War Plans Division. Tasked with estimating air force production requirements for conducting war against Germany and Japan, George mustered former ACTS colleagues for the initial planning effort. Employing the same planning methodology they had taught at ACTS, within nine days they had produced AWPD-1 (appendix 2), the USAAF's plan for defeating the enemy on both fronts. AWPD-1 formed the blueprint for aircraft production requirements designed to make HADPB central to the USAAF war effort.

The reality of air warfare in World War II soon revealed four flaws in the assumptions of HADPB when the theory was put into practice. First, formations of self-defending bombers could not, in fact, conduct unescorted offensive operations with acceptable losses against an integrated air defense system. A fleet of bombers could not simply avoid enemy forces by overflying the front, as strategic bombing advocates had claimed. While the static trench warfare of World War I may have been avoided, attritional battles were still being waged, this time in the skies.[2]

Second, HADPB required good weather for precision strikes with the Norden bombsight. The device proved unusable for most of the year, with the exception of Germany's sunny short summer. The USAAF lost its window of opportunity to test HADPB in the summer of 1943. Too few bombers arrived to Eighth Air Force in England, the result of aircraft production delays and the diversion of aircraft to the Pacific and Mediterranean theaters.[3] Eighth Air Force subsequently lost the summer of 1944 when ordered to interdict German transportation lines in France in support of the long-anticipated Allied invasion of Europe.

Third, the assumption that a wartime economy would be paralyzed by surgical strikes against a select set of vital and vulnerable nodes proved illusive. The resilient German economy was not as fragile as anticipated, in part because the Germans did not fully mobilize until 1943. Under the constant threat of day and night air raids, the Nazis prioritized and dispersed production facilities to maintain output. Economic collapse did not occur until the winter of 1944–1945. Rather than the culmination of select strikes against key economic nodes as anticipated by HADPB, however, the collapse was brought on by the systematic destruction of the entire German transportation network.

Fourth, when the German economy finally did break down, it did not produce the decisive results ACTS theorists had expected. The Nazi regime maintained strict control over the population and, even when faced with economic ruin, the Germans continued to fight to the end, unwilling to accede to unconditional surrender. Not until its army had been defeated and when the nation was faced with occupation by the Soviet Red Army did what was left of the Nazi regime finally agree to peace in May 1945.

Though US airmen fell short of independently achieving decisive victory, this did not mean that HADPB was inconsequential to the war's outcome. First, HADPB provided the prewar rationale to construct the thousands of B-17, B-24, and B-29 heavy bombers that offered the United States the offensive capability to conduct air campaigns against Germany and Japan. Further, offensive daylight air operations by Eighth Air Force in early 1944 were key to the USAAF gaining air superiority over western Europe, a prerequisite

High-Altitude Daylight Precision Bombing in World War II 197

for Operation Overlord, the invasion of western Europe. Bomber formations protected by long-range escorts forced the Luftwaffe into the sky against great odds to fight and die to protect the Fatherland. Though command of the air would continue to be contested throughout 1944, by June, USAAF fighters had cleared the skies above the beaches of Normandy. Eighth Air Force bombers further contributed to the successful effort to interdict German reinforcements. Finally, HADPB raids against German petroleum production facilities starved the German war machine of fuel such that, by December 1944, the German Army no longer had the means to maintain its desperate counteroffensive at the Battle of the Bulge.

Air power did prove critical to the Allied victory, though it did not do so in the independent and decisive way envisioned by ACTS. Air power was a key component of a combined arms strategy. This chapter examines the role American strategic bombing played during World War, beginning by sketching out the early air war in Europe leading up to the entry of the United States into the conflict. The development of US air war plans are considered next, followed by an assessment of the Combined Bomber Offensive (CBO), evaluating how well the underlying assumptions of HADPB withstood the harsh reality of war. Air power is then examined in its overall contribution to the Allied victory in Europe. Finally, the chapter concludes with USAAF experience with HADPB against Japan.

AIR WAR IN EUROPE, 1939–1941

Britain was ill prepared for war in September 1939 when Germany invaded Poland. Fearful of retaliatory attacks and equipped with a limited number of twin-engine bombers, Bomber Command restricted operations to coastal raids and leaflet drops over German cities during the "Phoney War."[4] But even these tasks proved too arduous given the difficulties encountered with poor flying weather, the inability to conduct long-range navigation, and the daylight threat from enemy fighters and anti-aircraft fire.[5] Forced to prioritize aircraft survival over lethality, Bomber Command chose to fly at night. Though night tactics neutralized German fighters and anti-aircraft fire, these advantages came with even greater challenges to navigation, target identification, and bombing accuracy.

The German invasion of the Low Countries in May 1940 brought an end to the Phoney War, and the Luftwaffe's carpet bombing of Rotterdam provided a foretaste of the devastation that would be wrought throughout Europe by aerial attack. Then, after the fall of France, the Luftwaffe turned its attention to the British Isles. Here, however, the Germans were met with a determined resistance from the RAF. Its new integrated air defense system was equipped with early-warning radar and quick-reaction fighter interceptors capable of concentrating against the incoming bombers. When German bombs inadvertently struck east London in late August, Prime Minister Churchill ordered Bomber Command to respond in kind against Berlin. Hitler retaliated with strikes on London. The Luftwaffe lost the Battle of Britain as it abandoned efforts to achieve air superiority and instead focused on nighttime raids on British cities.[6]

Though the Luftwaffe failed to bomb the British into submission, the RAF's Bomber Command remained committed to the strategic bombing of Germany. The RAF believed that the failure of German strategic bombing lay in two factors: first, in the inferiority of German twin-engine bombers (as compared to the new four-engine Lancaster bombers under construction by the British) and second, in the resilience of the British population (as compared to the fragile nature of the Germans, evidenced by the collapse of German morale at the end of World War I).[7]

Bomber Command initially prioritized the targeting of German oil production but poor weather, continued troubles with navigation, and the inability to hit point targets led to calls within the RAF to turn to area bombing. In May 1941, retired air marshal Sir Hugh Trenchard entered the debate by calling for the dispersed area bombing of industrial cities to undermine the morale and productivity of the German workers. By late summer, however, the infamous Butt Report assessed Bomber Command as incapable of even locating and hitting targets the size of cities.[8]

Despite the fact that heavy air-to-air losses had compelled both the Germans and the British to switch to night bombing, the American airmen who observed the air war remained resolute in their doctrine of high-altitude daylight precision bombing.

Air War Plans Division

In July 1941, General Hap Arnold, the chief of the USAAF, summoned Lieutenant Colonel Harold George to Washington to set up the Air War Plans Division.[9] Simultaneously, President Roosevelt directed the army and navy to

US Army Air Staff, 1941: Lieutenant Colonel Harold George (*second from left*), Major General Hap Arnold (*center*), and Major Haywood Hansell (*third from right*).

provide estimates on the production requirements for conducting war against Germany and Japan. At the end of July, the task to develop and integrate air force requirements into the overall army plan fell to AWPD, which by then consisted of George and a small staff including a core cadre of trusted former ACTS faculty: Kenneth Walker, Haywood Hansell, and Laurence Kuter.[10] As the Bomber Mafia machinated, General Arnold departed for Newfoundland to accompany President Roosevelt in talks with the British prime minister Winston Churchill. AWPD had only nine days to assemble a scheme before Arnold's return on August 12.[11]

General Arnold gave AWPD few instructions other than to develop a plan that focused on defeating Germany first while maintaining a defensive position in the Pacific. George and his cohort were determined to place an independent strategic bombing campaign at the center of the American war effort. Still, in order to meet with army staff approval, they included fighters and attack aircraft in the event a ground invasion proved necessary.[12]

To calculate the overall aircraft and manning requirements, AWPD first had to decide how the air forces would be employed. In forming their strategic bombing campaign plan, they first identified the vital and vulnerable targets that, if destroyed, would paralyze the German war economy. Haywood

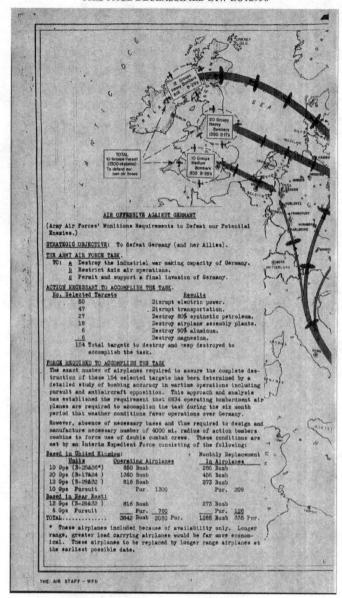

AWPD-1: Air Offensive against Germany, August 1941. The strategic objective of the offensive was the defeat of Germany, with the Army Air Force tasked with destroying the German industrial war making capacity, restricting Axis air operations, and permitting and supporting a final invasion of Germany. Specifically, the goal was the destruction of 154 targets with the intent to "disrupt electric power," "disrupt transportation," "destroy 80% synthetic petroleum," "destroy airplane assembly plants," "destroy 90% alu-

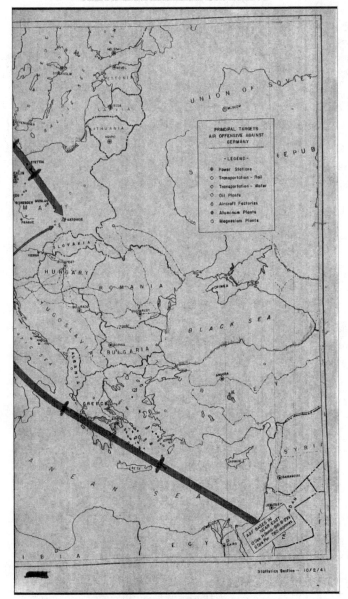

minum," and "destroy magnesium." Requirements for the offensive included 6,834 bombers and a six-month period with favorable weather conditions over Germany. Due to the unavailability of a 4,000-mile radius bomber, an interim expedient force consisting of 3,842 bombers and 2,080 pursuit aircraft would be based in the United Kingdom, with a monthly replacement of 1,288 bombers and 335 pursuit.

Hansell was responsible for this task. Since 1940, he had been assigned to the Air Staff's Strategic Air Intelligence Office, where he studied the German economy and, with the assistance of civilian specialists, had prepared target lists of German petroleum and electrical power plants. In early 1941 Hansell traveled to London to share intelligence with RAF planners who, in turn, provided valuable information on German aircraft production, transportation networks, and the Luftwaffe organization. From these studies, Hansell identified 154 primary targets under the categories of electric power, transportation, synthetic petroleum, airplane assembly plants, and the major sources of aluminum and magnesium (see AWPD-1 in appendix 2).[13]

With the target set identified, it was up to Laurence Kuter, who had taught Bombing Probabilities at ACTS, to calculate the number of bombs required to destroy the target list. Informed by Bomber Command's combat experience, the bomb total calculated from peacetime bombing charts was multiplied by a factor of two and a quarter. Factoring in likely combat losses, mission aborts, and European weather patterns, which on average allowed only five clear days for bombing per month, the planners estimated the total number of bombs and bombers required for the air operations.[14]

Surprisingly, the army staff and the Roosevelt administration accepted the air plan, named AWPD-1, with little modification. It ambitiously estimated an air force of nearly 2.2 million men and 63,000 aircraft, 7,000 of which would be heavy bombers dedicated to the air offensive against Germany. The planners expected to need eighteen months to generate the requisite aircraft and aircrews and another six months to conduct the air offensive needed to independently win the war.[15]

In December 1941, following the attack on Pearl Harbor, additional air forces were added to the plans for the Pacific, but the overall scheme remained intact. Roosevelt requested an update on air force requirements in August 1942.[16] Hansell, the last of the ACTS cadre still in Washington, led the reassessment. AWPD-42 (appendix 3) did not depart from the methodology employed by AWPD-1. The revised plan called for an army air force of 2.7 million personnel and 75,000 aircraft by 1943, with the increase in the number of aircraft reflecting training requirements. The overall strategy focused on Germany first. The priority of the USAAF was to "deplete the German Air Force, destroy the sources of German submarine construction and undermine the German war-making capacity." The emphasis placed on striking the Luftwaffe was a recognition that gaining the level of air superiority needed to conduct a daylight strategic bombing campaign, or to conduct an invasion of the Continent if strategic bombing failed to be decisive, was going to be more difficult than ACTS theorists had previously anticipated. The inclusion of submarine facilities in the target list pointed to the urgency of the Battle of the Atlantic at the time and acknowledged that the air campaign depended on keeping Britain in the war so that American bombers could launch from English bases.

THE COMBINED BOMBER OFFENSIVE

In February 1942, Air Chief Marshal Sir Arthur Harris assumed command of Bomber Command and prioritized the nighttime targeting of German industrial cities. By the end of May, Bomber Command had launched its first thousand-bomber raid against Cologne, with more large-scale night raids against industrial cities in western Germany to follow.[17]

US forces assigned to Eighth Air Force began arriving in England in the summer of 1942. The American airmen were unwilling to join the British in the nighttime area bombing of German cities, instead insisting on daylight raids against specific targets. The numbers of aircraft available to Eighth Air Force remained insufficient to conduct large air raids that year, however, as US bomber production lagged behind requirements and many of the aircraft that did roll off assembly lines were diverted to North Africa and the Pacific.[18]

At Casablanca in January 1943, the American and British Combined Chiefs of Staff settled on a Combined Bomber Offensive against Germany (see appendix 4), which did little more than codify the divergent strategic bombing campaigns already being waged by the two allies. The stated objective was "the progressive destruction and dislocation of the German military, industrial and economic system, and the undermining of the morale of the German people to a point where their capacity for armed resistance is fatally weakened." The statement captured both the USAAF aim to attack the German war economy and Bomber Command's objective of targeting the German population. The target list prioritized submarine construction yards, aircraft industry, transportation, oil plants, and other targets of the enemy war industry. The memorandum further encouraged both day and night bombing to maintain pressure on German morale, to attrit the Luftwaffe fighter force and draw fighters away from the Russian and Mediterranean theaters, to support operations in the Mediterranean, and to prepare for the invasion of the Continent.[19]

British bombing in the spring of 1943 focused on the Germans' industrial Ruhr Valley, and by the summer of 1943, airstrikes had capped the expansion of Nazi war production. A major turning point in the Combined Bomber Offensive, however, came in late July 1943 with the bombing of Hamburg. A dry, hot summer had set conditions for a major firestorm when incendiary bombs struck Hamburg. The flames engulfed the city, killing and injuring a staggering 80,000 inhabitants, destroying a quarter of a million homes, and leaving 900,000 homeless. The operational success of the Hamburg raid encouraged Bomber Command to turn its attention further east to launch massive raids against Berlin in November in an effort to knock the German capital out of the war.[20]

Testing High-Altitude Daylight Precision Bombing

Like Bomber Command, after Hamburg Eighth Air Force conducted strikes progressively deeper into Germany. Unlike Bomber Command, which intentionally targeted cities, Eighth Air Force aimed at fighter aircraft factories and ball bearing plants in accordance with the Pointblank Directive issued in June 1943, which prioritized the attainment of air superiority in western Europe (appendix 5).[21] In a series of air raids in the late summer and fall of 1943, Eighth Air Force successfully attacked factories in central Germany, but in the process suffered defeat as its bombers were overwhelmed by Luftwaffe fighters. On August 17, 60 of 376 bombers were lost (16 percent attrition) over Regensburg and Schweinfurt, and on September 6, 45 out of 338 bombers (13.3 percent) didn't return from a raid on Stuttgart. In a third raid, reattacking Schweinfurt on October 14, Eighth Air Force sacrificed 60 of 291 B-17s at an unsustainable attrition rate of 20.7 percent.[22] These horrific losses curtailed US efforts to conduct unescorted high-altitude daylight precision bombing against Germany. It would not be until February 1944, with the introduction of long-range fighter escorts, that USAAF bombers would reappear in force over Germany.

The Eighth Air Force defeat in the autumn of 1943 dispelled the notion of unescorted daylight bombing. By March 1944, improvements in German night fighter tactics further removed the sanctuary of darkness for Bomber Command, which suffered combat losses that matched those of Eighth Air Force in the previous fall.[23] The German integrated air defense system, with its deadly combination of early-warning radars and quick-reacting fighter

B-17F formation over Schweinfurt, August 17, 1943.

interceptors, disproved the prewar assertions of ACTS that a well-armed formation of unescorted bombers could conduct long-range operations at an acceptable loss rate.

The bloody air war over Germany dispelled another belief of the strategic bombing advocates: that air power had obviated the need for attritional battle. As the Combined Bomber Offensive continued and losses mounted, it became obvious that the bloody trench warfare of World War I had simply been elevated into the sky. As with surface forces, air forces had to first achieve the intermediate objective of defeating enemy forces before direct pressure could be placed upon the enemy nation. As it turned out, strategic bombing did not end wars quickly with fewer overall casualties, as the proponents of strategic bombing had promised.[24]

Another implicit assumption by ACTS that proved untenable was the feasibility of conducting a high-altitude daylight precision bombing campaign against Germany. In practice, Eighth Air Force could not achieve pin-

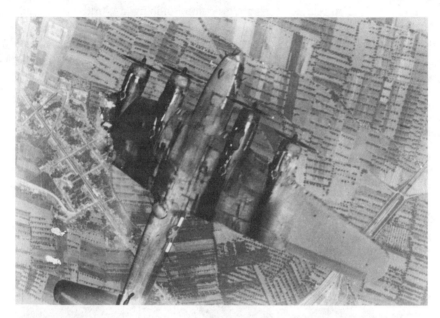

B-17F inverted over Germany after losing right wing, circa 1943.

B-17 formation dropping bombs through clouds over Germany in 1945. (National Archives)

point accuracy due to German defenses and the weather. The threat from enemy fighters forced Eighth Air Force to abandon the tactic of having each bombardier calculate the bomb release point for his own aircraft. Instead, in order to maintain a tight defensive formation, only the lead bombardier calculated the bomb release point and the other bombardiers dropped on command. This tactical innovation to reduce the vulnerability of the bombers to enemy fighters diminished bombing accuracy by one-third. Other German defensive measures such as concentrating anti-aircraft fire around key targets forced bombers to fly at higher altitude, a maneuver that further decreased the accuracy of the Norden bombsight. Passive defensive measures employing decoys, camouflage, and smoke screens further affected bombing accuracy by making target identification more difficult.[25]

Even more degrading to US bomber accuracy than the enemy's defenses was the impact of weather over England and Germany. Weather conditions limited the number of days bombers could launch from England and find unobscured skies over Germany. Eighth Air Force's weather abort rate was only 10.4 percent but, of the missions launched, only half were able to bomb visually. Daytime weather conditions over Germany during the winter proved particularly onerous. By November 1943, Eighth Air Force effectively abandoned the pretense of precision bombing when ordered by General Arnold to bomb through overcast conditions if need be. Eighth Air Force adopted radar bombing techniques that in practice resulted in area bombing and proved to be even less accurate than the night tactics adopted by Bomber Command.[26]

HADPB could not be conducted through cloud cover, a limitation that prevented the systematic attack of the vital economic nodes that the ACTS theorists believed would paralyze the German war economy. Only during the summer months was the weather dependable enough to attempt a high-altitude visual bombing campaign, but the United States squandered its first opportunity in the summer of 1943. A combination of delayed aircraft production and the dispersion of aircraft to other theaters limited the number of bombers available to Eighth Air Force.[27] But even if Eighth Air Force had had more bombers in the fall of 1943, Germany's early-warning radar and integrated air defense system had by then so shifted the offense-defense balance in favor of the defense that daylight unescorted bombing raids would have proven unsustainable. The following summer of 1944 was likewise a lost opportunity for Eighth Air Force to test HADPB. It was at this point that

General Eisenhower took control of Allied targeting selection in preparation for the Allied invasion of Europe. Not until the end of September 1944 did Eighth Air Force regain its authority to select targets and refocus its attention on dismantling the German economy.

In addition to failing to anticipate the impact of enemy defenses and weather on their ability to conduct HADPB, ACTS theorists underestimated the resiliency of the German economy and population. In 1941, AWPD planners had assumed the German economy was operating at maximum capacity. This was not the case, however, as Germany had not yet fully mobilized and thus had the slack to absorb the impact of air strikes. It would require repeated strikes to keep factories offline. The Germans further anticipated Allied air strikes by dispersing production, which increased costs and decreased total potential output but did not significantly curtail production. Vital and vulnerable economic nodes proved more difficult to identify, attack, and reattack than anticipated. Not until the winter of 1944–1945 did airstrikes effectively dismantle the German transportation system, which then finally led to economic collapse.[28]

A final assumption of HADPB theory was that the dislocation of the wartime economy would deny the enemy the means to continue fighting and, paired with the economic suffering of the population, would compel Germany to sue for peace. ACTS theorists, however, had underestimated the resolve of the German population, or at least the strength of the Nazi regime to control its citizens. The collapse of the economy did not, in fact, bring about an end to the war as ACTS theorists had expected, nor did it obviate the necessity to invade.

In addition to Nazi domination over its population, one reason the Germans were unwilling to agree to a peace deal arose from Roosevelt's decision, announced at the Casablanca Conference in January 1943, to demand unconditional surrender.[29] There was little incentive for the Nazis to concede to such a dire ultimatum while they still had a hope of victory, or at least believed military resistance might modify Allied demands.

What High-Altitude Daylight Precision Bombing Achieved

The previous section considered the reasons why high-altitude precision bombing failed to win the war independently. HADPB was ineffec-

High-Altitude Daylight Precision Bombing in World War II   209

tive because of the combination of lethal enemy air defenses, cloud cover, a resilient German war economy, and a brutal Nazi regime that controlled its population. Although the USAAF did not achieve the decisive victory in the manner that ACTS theorists had anticipated, HADPB still made four significant contributions to the Allied victory: it provided the prewar rationale for an offensive bomber force, played a critical role in obtaining Allied air superiority in Europe, seriously reduced German synthetic fuel production, which was indispensable to the German military, and helped interdict the German lines of communication in France.

HADPB as Prewar Rationale for Long-Range Bomber Force

First, HADPB provided the prewar argument for an independent, offensive-minded air force. The US air planners requested thousands of long-range, heavy bombers in order to destroy the vital and vulnerable economic targets identified in AWPD-1. The acceptance of AWPD-1 by President Roosevelt in late 1941 initiated the production of the B-17s and B-24s that arrived monthly by the hundreds in England and the Mediterranean in late 1943 and early 1944. Without the ACTS theory as a blueprint, the United States would not have generated the capacity to conduct a long-range air offensive.[30]

During the interwar period, only British and American airmen made strategic bombing central to their respective air power doctrines.[31] Both prioritized the development of four-engine bombers capable of long ranges and carrying heavy bomb loads. By contrast, the Luftwaffe, which adopted a combined arms doctrine for the continental land wars it anticipated, had only a medium-range bomber force at its disposal. Although the United States did not have an air force in being at the outset of the war, the Air Corps' interwar technological development of the B-17 enabled the United States to quickly generate an offensive bomber fleet.

Attaining Air Superiority over Europe

The daytime bombing of Germany played a crucial role in the United States gaining air superiority over Europe. Luftwaffe fighter attrition had been high but bearable since the invasion of France in 1940. In early 1944, however, the losses proved unsustainable when Eighth Air Force bombers returned in mass over Germany accompanied by lethal long-range fighter escorts.[32] German fighter losses, which had hovered at 30 percent per month during the fall and winter of 1943, spiked in March 1944 to 56.4 percent and re-

Republic P-47 Thunderbolt with external drop tank, circa 1944.

mained between 43 percent and 50 percent through the Normandy landings in June.[33]

Though it was USAAF fighters that inflicted the deathblow on the Luftwaffe, it would be a mistake to consider Eighth Air Force bombers immaterial to the attainment of air superiority for the Allies. Prior efforts by the RAF to lure the Luftwaffe into the skies against similarly unfavorable odds had failed, as witnessed by the "Circus" daylight raids over northern France in 1940–1941. The difference in 1944 was that Eighth Air Force had the capability to penetrate deep and strike at the heart of Germany in the daylight. Consequently, these raids compelled Luftwaffe fighter pilots into the sky, under unfavorable odds, to fight and die.

As a result of US raids on Germany, by June 6, 1944, the skies over the beaches of Normandy were clear and, while Allied forces faced many challenges conducting their amphibious landings, enemy air attack was not one of them. Because the USAAF had the foresight to generate a daylight long-range bombing capability, and the later flexibility to add long-range fighters,

North American P-51 Mustang formation, circa 1944.

the United States was able to gain air superiority, a prerequisite to the invasion of western Europe.

HADPB and German Synthetic Fuel Production

During the summer of 1944 HADPB strangled German synthetic fuel production, the lifeblood for the German Army and Air Force.[34] Precise daylight strikes were required to achieve the accuracy and bomb density necessary to disrupt the petroleum plants. The Germans concentrated roughly half of their synthetic oil production at four sites, and on two clear days in May 1944, Eighth Air Force successfully struck these and five other facilities. Follow-up HADPB raids throughout the summer markedly reduced German finished oil production by 42.6 percent. The loss of high-grade fuel production severely restricted Luftwaffe operations and delayed a German counteroffensive. The return of cloud cover in October and November prevented the continuation of HADPB strikes, however, and the Germans recommenced synthetic fuel production to support limited air operations and generate sufficient stores to launch, though not sustain, the German counteroffensive in December 1944 at the Battle of the Bulge.[35]

Interdiction of German Forces

Finally, the USAAF interdicted German lines of communication in France and western Germany in preparation for the Normandy invasion in June 1944. The Germans had the advantage of interior lines between the Eastern and Western Fronts, which in theory meant they could more easily redeploy their forces between the two theaters. The disruption of the German rail transport, however, delayed the deployment of desperately needed reserves to the west. In late April, Eighth Air Force diverted its bombers to attack the transportation network in France, Belgium, and western Germany. B-17 bombers participated in the interdiction campaign, conducting visual strikes against marshalling yards when the weather permitted and employing their H2X radar bombing system when skies were overcast. Radar bombing meant in practice the area bombing of the villages and towns where the marshalling yards were located. Simultaneously, Allied fighters and medium-range bombers conducted more precise low-altitude attacks against rail and road bridges and destroyed rolling stock.[36]

The impact of the Allied interdiction campaign was exacerbated by the Germans' decision to keep a large number of its ground forces in reserve. In the spring of 1944 the Germans anticipated the Allied invasion, but its army leadership was split over how best to defend France. The view held by the commander in chief of the west, General Karl von Rundstedt, was to keep panzer tank units as a strategic reserve to be able to quickly reinforce the German lines wherever the Allies should strike. He thus chose not to disperse all of his forces along the coast, where a concentrated Allied offensive could overpower the Germans at any given point.[37]

Rundstedt's strategy made sense, however, only if the Americans had not had air superiority and Allied aircraft had not been capable of delaying his reserves. General Erwin Rommel disagreed with Rundstedt and lobbied Hitler for the forward defense of France. Rommel had learned from the North African campaign that the Germans would not be able to maneuver their armored units under the threat of Allied air power. He unsuccessfully argued that the best chance for the Germans was to use all ground forces available to fortify the coast.

When the invasion came, German reserves were seriously delayed by the disruptions in railways and roads and being forced to move at night or under cloud cover to avoid attack by US fighters.

Alternative Explanations for the Utility of HADPB

Though obtaining air superiority, restricting oil production, and interdicting German lines of communication played a role in the Allied victory, these contributions fell well short of the independent and decisive results predicted by the ACTS theorists. After the war, strategic bombing advocates argued that HADPB had affected the outcome of the war in additional ways: keeping the Soviet Union in the war, disrupting the German war economy, diverting equipment and personnel from the Eastern Front, and compelling the Germans to prioritize the production of fighter aircraft, anti-aircraft artillery, and revenge weapons.[38] These ex post claims are largely inaccurate. Each argument will be analyzed in turn.

Keeping the Soviet Union in the War

One argument on the efficacy of strategic bombing is that the Combined Bomber Offensive maintained the cohesion of the Grand Alliance at a critical juncture by dissuading Stalin from making a peace deal with Hitler.[39] Roosevelt had promised a second front on the Continent but by the summer of 1942, Churchill and Roosevelt had concluded that an invasion was not feasible until 1943 at the earliest.[40] Instead, the US Army would first join the British Army in the North African campaign and the USAAF would join Bomber Command in England for the CBO. In August 1942, Churchill traveled to Moscow to break the news to Stalin.[41] To cushion the blow, Churchill offered a strategic bombing campaign as an alternative to a second front, arguing that the bombing of cities and attacking wartime industries would weaken Germany and compel Hitler to divert men and equipment from the Eastern Front to defend the homeland. Personally, Churchill was skeptical of the efficacy of strategic bombing since the Butt Report the previous year had exposed the ineptness of Bomber Command. Still, strategic bombing provided Churchill the rhetoric to use with Stalin, and with the British public: that the British and Americans were taking the fight directly to Germany.

The key to the belief that the CBO kept the Soviet Union in the war hinges not only on the question of whether Stalin was mollified by the British and American strategic bombing campaigns but also on the matter of whether he ever had the opportunity to make a peace deal with Hitler. Stalin, in fact, was never satisfied with the CBO as a substitute for a second front, which remained the Soviet's top priority for the Grand Alliance.[42] But besides voicing his contempt for the continual delay in the western invasion, Stalin had little choice but to maintain the alliance. So long as Germany remained on the offensive in the east, which it did through the Battle of Stalingrad that ended in February of 1943, Stalin's only options were to keep fighting or surrender. After the German defeat at Stalingrad in February until the Battle of Kursk in July, however, the shift in the balance of power did offer a window of opportunity for a peace deal, had both sides been so inclined.[43]

In 1942 and 1943, the Soviets made peace overtures through informal diplomatic channels in Sweden, but Germany continually rebuffed any rapprochement.[44] In May 1943, Stalin abandoned these clandestine efforts and publicly demanded unconditional surrender. In June, however, Roosevelt and Churchill informed him that the invasion of France would be again delayed until 1944, a setback that caused Stalin to reverse course and reattempt to open diplomatic channels with the Germans at Stockholm.[45] The CBO did not dissuade the Soviets from reaching out to the Germans despite the fact that, by June 1943, Bomber Command had finally made a significant impact on the German economy with its ongoing Battle of the Ruhr. If ever the CBO could have convinced Stalin of its efficacy, it was then. Based on the available evidence, however, it was more likely Hitler's intransigence rather than the CBO that kept Stalin in the war. Any last window of opportunity for a peace deal slammed shut with Germany's decision to commence the Battle of Kursk in early July in an effort to preempt the anticipated Soviet counteroffensive.[46]

Impact on German War Production

A second potential justification for the utility of HADPB was its impact on the German economy. For decades, critics of strategic bombing have argued that the CBO had little impact on German war production.[47] Adam Tooze, however, in his revisionist work *Wages of Destruction: The Making and Breaking of the Nazi Economy*, has demonstrated that the CBO halted the rapid expansion of German industrial production.[48] The bombing of the Ruhr in the spring of 1943 disrupted steel production. As a result, the Germans had severe shortages for the duration of the war.

The problem with this argument in support of HADPB is that it was Bomber Command's night raids, and not the USAAF's daylight precision strikes, that capped German industrial expansion. From March to June 1943, Bomber Command conducted repeated raids on key industrial cities in western Germany that proved much more effective than prior raids due to three factors.⁴⁹ First, the RAF had responded to navigational and targeting deficiencies exposed by the Butt Report by developing electronic navigation aids, the first of which was Oboe, a radio transponder system capable of the accurate blind bombing of targets as far away as western Germany.⁵⁰ Second, the introduction of the Mosquito light bomber pathfinder force further increased the accuracy of nighttime raids by utilizing master bombardiers skilled in the use of air-to-ground radar, who could identify and mark targets for the incoming bomber stream.⁵¹ Third, the rise in both the quantity and quality of the heavy bomber aircraft available to Bomber Command increased the lethality of its air strikes. The Lancaster had a much larger bomb load than earlier bombers, and by the spring of 1943 over one hundred were rolling off assembly lines each month.⁵²

By contrast, in the spring of 1943 USAAF did not yet have the forces available to mount serious raids on Germany. It would not be until late summer that Eighth Air Force had sufficient bombers and crews for large-scale attacks, but by then the Pointblank Directive (appendix 5) had prioritized the obtainment of air superiority over the destruction of the German economy. As a result Eighth Air Force focused on attacking German fighter aircraft and ball bearing production and not on intermediary industrial goods.⁵³

It would not be until a year later, in late September 1944, that Eighth Air Force again prioritized the German economy by targeting its transportation network. However, by this time clouds covered much of Germany, preventing visual bombing with the Norden bombsight for 90 percent of the time.⁵⁴ Eighth Air Force was again compelled to use its H2X air-to-ground radar to area-bomb villages and towns along the major rail lines. The targeting of German transportation finally caused the collapse of the German economy over the winter of 1944 and 1945. Rather than the result of HADPB, it was brought about by a combination of area bombing of the marshalling yards and the destruction of bridges and rolling stock wrought by fighter and medium bombers with low-altitude attacks.⁵⁵

Diverting German War Resources

A third argument for the contribution of HADPB to the Allied war effort is that Eighth Air Force operations compelled Germany to divert fighter aircraft, anti-aircraft artillery (AAA, or flak guns), and personnel away from the Eastern Front. Eighth Air Force daylight bombing raids were, in fact, responsible for the German decision to rotate fighters home to defend the Fatherland. Unfortunately, the impact proved too little and too late to relieve the Soviets when they needed it the most, from the German invasion in June 1941 through the termination of the German offensive following Stalingrad in February 1943. The Luftwaffe did not begin to divert a significant number of fighters for homeland defense until after the Battle of Kursk in July. Whereas there were twice the fighters on foreign soil as in Germany in June, by October the ratio was one-to-one.[56] At this point, however, the Soviets had withstood the full brunt of the German invasion and were transitioning from defense to offense. Though additional Luftwaffe fighters would have been useful against the Soviet offensive, given the growing advantage in Soviet power, the most German air power could have been expected to achieve was to slow the Red Army advance.[57]

As for the German decision to divert flak guns, it was Bomber Command, not Eighth Air Force, that was responsible. The Germans reinforced their defenses from 800 heavy flak batteries in 1940 to over 2,100 by 1943. By this time flak guns made up a quarter of all German weapons production and consumed 17 percent of German ammunition.[58] The 36,000 heavy and light AAA pieces defending the homeland were double the surface-to-air weapons deployed in other theaters.[59] By early 1943 Bomber Command had expanded to be able to conduct 1,000 bomber raids while Eighth Air Force was still struggling to get 100 aircraft airborne.[60] The Soviets indeed faced fewer guns on the Eastern Front than they otherwise would have, but this was thanks to Bomber Command's nighttime raids and not Eighth Air Force's high-altitude daylight precision bombing.

In addition to the diversion of fighters and flak guns, nearly a million personnel were required to operate the German air defense system. This large manpower reserve could, in theory, have been made available for duty on the Eastern Front. In practice, however, the Germans diverted factory workers, many deemed unsuitable for soldier duties, to operate the air defense system. Those workers capable of military service manned the anti-aircraft guns which, as just argued, were primarily intended to defend against Bomber Command.[61]

Reprioritization of German Aircraft Production and Revenge Weapons

A final argument for the positive effect of HADPB was that it caused Germany to shift emphasis toward defensive fighter aircraft production and divert scarce resources for the research and production of the V-1 and V-2 missile systems. In terms of the reprioritization of German aircraft production, through 1942 the Nazis had focused more on offensive operations, producing one and a quarter bombers for every fighter. In 1943, however, the ratio was inverted, with one and a third fighters being built for every bomber, and by 1944 fighters were built exclusively. This shift in production had less to do with the threat from HADPB, however, and reflects more the overall shift by Germany to a defensive strategy. In early 1943, the Luftwaffe had largely abandoned air support for its armies, a decision made after its poor showing in the Battle of Stalingrad.[62]

With regard to the German vengeance weapons, the decision to produce the V-1 cruise missile and V-2 ballistic missile was unaffected by the USAAF air campaign. The German Army had been investing heavily in the V-2 since its Peenemunde missile facility opened in 1937. Priority for the development of a ballistic rocket remained high thereafter. The Luftwaffe began to develop the V-1 cruise missile in the spring of 1942 as a response to British attacks on German cities. By the winter of 1942, with Germany's strategic position deteriorating, Hitler looked to the V-1 and V-2 as a retaliatory response to Bomber Command's nighttime raids. Due to the significant technological challenges in designing a cruise missile, and even more so a ballistic missile, the V-1 was not operational until June of 1944 and the V-2 not until September. These programs drained significant resources from other German weapons programs and netted little for the German war effort, but Hitler's decision to produce these weapons had nothing to do with the American strategic bombing effort.[63]

HIGH-ALTITUDE DAYLIGHT PRECISION BOMBING AGAINST JAPAN

Unlike in Europe, where B-17s and B-24s had the range to reach most targets, the extreme distances of the Pacific required the USAAF to employ its newest long-range bomber, the B-29. In 1939, the Air Corps initiated plans to develop a bomber with the ability to fly from coast to coast. Due to a series of technological challenges, however, the first prototype would not fly until 1942. Delays associated with the complexity of the aircraft design and enduring problems with its high-performance engines plagued the bomber's development and production schedule.[64] When the B-29 did at last come online,

Boeing B-29 Superfortress, circa 1945.

however, it's pressurized cockpit allowed flight above 30,000 feet and, with a top speed of 350 miles per hour, the B-29 proved difficult for Japanese fighters to intercept. Further, the B-29's combat range of 1,600 miles doubled that of the B-17, with triple the bomb load, to boot.[65] The leap in service ceiling, speed, range, and bomb load came at a high price, however, as the program ended up costing $3 billion, making it the most expensive weapon system of the war, well above the $2.1 billion spent on the Manhattan Project.[66] With such an enormous outlay of resources, the USAAF desperately needed the new bomber to play a major role in the defeat of Japan. In light of the indecisive results of strategic bombing in Europe, airmen's dreams of an independent air force now rested with the B-29.

Unfortunately, the USAAF did not similarly invest in a new precision bombsight. Like its predecessors, the B-29 utilized the Norden, which proved inadequate for the HADPB of Japan for three reasons. First, similar to weather patterns in western Europe, clouds frequently obscured the Japanese isles. While the B-29 was equipped with the APQ-13 radar as a weather backup, similar to the H2X on the B-17, the air-to-ground radar was suitable only for area bombing. Second, Norden's telescopic sight was not designed for the extreme slant ranges confronted when dropping bombs from above 30,000 feet. To hit their targets, aircrew were compelled to descend to lower altitude, making them susceptible to enemy flak and fighters. Third, the jet stream encountered over Japan was much stronger and at lower altitudes than that over Europe, with winds frequently above 120 miles per hour and exceeding both the limits of the Norden bombsight and the capability of even the most seasoned bombardier.[67]

To avoid being diverted to other priorities by theater commanders, as had happened in Europe when bombers were diverted to support the North Africa Campaign, all B-29s were assigned to 20th Air Force, headquartered in Washington, DC, and commanded by the chief of the USAAF, General Hap Arnold. Arnold planned to base the B-29s in China (XX Bomber Command) and the Marianas Islands (XXI Bomber Command). The Committee for Operations Analysts (COA) had commenced initial target selection for a strategic bombing campaign against Japan the previous year and, by the winter of 1943, it had identified, but not prioritized, six target sets: steel plants, merchant shipping, aircraft factories, ball bearing plants, radar and radio facilities, and urban industrial areas. Steel production, as an intermediate good, was the only target that fit neatly into the ACTS strategic bombing theory for targeting vital and vulnerable economic nodes. By contrast, the attack of merchant shipping complemented the navy's ongoing blockade, and the targeting of aircraft and ball bearing factories, along with radar and radio facilities, supported an air superiority campaign, a prerequisite to ground invasion. The targeting of urban industrial areas was reminiscent of RAF Bomber Command's campaign against Germany, with the firebombing of Hamburg in July 1943 serving as a template for inflicting extensive damage upon a city. The COA considered Japanese cities particularly vulnerable to incendiary raids, given the large number of wooden structures.[68]

At the Cairo Conference in November 1943, Roosevelt promised Chiang Kai-shek that B-29s would soon be bombing Japan from China. Sustained operations from China, however, proved logistically and operationally infeasible. In 1942, the Japanese had invaded Burma, interdicting the Allies' only land route to China. Logistical support from India required that supplies be flown over the Himalayas, a herculean task involving seven flights over the "Hump" to ferry the fuel and bombs for a single B-29 combat sortie. Operationally, the distance from its base at Chengdu in south-central China limited the range of the B-29 to Japan's southernmost island of Kyushu.[69]

Despite all of these formidable challenges, 20th Air Force tried to make good on Roosevelt's promise. On June 15, 1944, the XX Bomber Command launched its first raid with a night attack on the steel works at Kyushu. Forty-seven of the 67 bombers dropped nearly 2,000 bombs, resulting in only a single hit on the complex. This came at the loss of seven aircraft due to pilot error or mechanical failure, while only a single plane fell to enemy flak. Subsequent raids throughout the summer were similarly unsuccessful. In late August, an impatient Arnold sent Major General Curtis LeMay to China to wield the combat experience he had gained in Europe in pursuit of better results. LeMay set about improving the training and tactics of the B-29 aircrew, but even he could not overcome the logistical challenges and range limitations of conducting operations from China.[70]

General Arnold's hopes for the B-29 turned to the Marianas Islands once the US Navy had secured Guam, Tinian, and Saipan in August 1944. Arnold selected Brigadier General Haywood Hansell, his chief of staff, to command XXI Bomber Command, which was deployed from Kansas to the Pacific. As Hansell was a former ACTS faculty member of the Bomber Mafia and key planner for both AWPD-1 and AWPD-42, it was unsurprising that he trained his aircrew on HADPB. On November 24, XXI Bomber Command launched its first mission against Japan, with 111 bombers attacking the Musashino aircraft plant on the outskirts of Tokyo. Results were poor. A mere 24 aircraft located the target, with only 48 of the 1,000 bombs released actually landing within the factory perimeter, inflicting little damage. The one positive note was that only one B-29 was lost to a Japanese fighter and another ditched due to fuel starvation. A follow-up raid on November 27 found Tokyo obscured, and subsequent raids through the end of December faced similar difficulties identifying targets. Even when factories were visible, B-29 bombardiers could not consistently hit their marks because of extreme winds and slant ranges.[71]

Washington continually pressed Hansell for better results. Brigadier General Lauris Norstad, who had replaced Hansell at the Pentagon and was effectively in charge of 20th Air Force while General Arnold recovered from a heart attack, repeatedly urged Hansell to conduct incendiary raids on Japanese industrial cities. Hansell resisted. Following the December 27 raid on Tokyo, regarding which Hansell acknowledged to the press the many operational problems yet to be solved, a recuperating and frustrated Arnold fired

Lauris Norstad, Curtis LeMay, and Thomas Power (*left to right*) on March 10, 1945, reviewing report on the firebombing of Tokyo. (National Archives)

Hansell. Curtis LeMay soon arrived from China and assumed command on January 20, 1945.[72]

For the next month, LeMay continued with the HADPB tactics with similarly meager results, all the while receiving continual prodding from Norstad to conduct incendiary raids. On February 25, 172 B-29s loaded with incendiary bombs employed their APQ-13 radars from high altitude to area-bomb Tokyo. Photographic evidence showed that the attacks had burned a square mile of the city. After two weeks of teetering between HADPB and area bombing, LeMay finally committed to the latter. On March 9, he ordered that the B-29s be stripped of their self-defense armament and flown at low altitude, measures that allowed for a doubling of the incendiary bomb load. The night attack on Tokyo proved to be the single most destructive air raid in history, killing more than 80,000 and destroying sixteen square miles of Tokyo. Afterward, XXI Bomber Command abandoned the pretext of HADPB, conducting four similar strikes over the next nine days that left much of Tokyo, Osaka, Kobe, and Nagoya destroyed. LeMay persisted with firebombing raids over the next five months until the war ended following the atomic bombing of Hiroshima and Nagasaki on August 6 and 9, 1945, respectively. The firebombing campaign destroyed the majority of Japan's industrial cities and, in the process, killed between 260,000 and 310,000 civilians, injured another 412,000, and destroyed 2.2 million houses, leaving 9.2 million people homeless.[73]

Why HADPB Failed to Independently Win the War against Japan

Allied victory was not the result of the independent air campaign envisioned by ACTS theorists, by which the Japanese economy would have been paralyzed by surgical attacks against its vital and vulnerable economic nodes. Surrender was secured, rather, by a combination of the naval blockade, the firebombing of Japanese cities, the Soviet invasion of Manchuria, and the atomic bombing of Hiroshima and Nagasaki.[74] B-29s contributed not only to the firebombing and atomic attacks but also to the blockade by the aerial mining of Japanese waters to interdict local shipping.[75]

In the end, the B-29's performance did provide an argument for an independent air force, though not as a result of HADPB. Aside from the employment challenges previously discussed (poor weather, high winds, and extreme slant ranges), there were three other reasons why HADPB failed. Foremost was target selection. The decision to prioritize aircraft factories made little sense by the end of 1944. The Japanese had already lost the fight for air superiority with their defeat in late October at Leyte Gulf. The attritional naval air battles with the US Navy from 1942 to 1944 had already given command of the air in the Pacific to American airmen. Further attacks on aircraft factories may have slowed production, but it was an insufficient number of experienced pilots, rather than a lack of aircraft, that plagued the Japanese war effort.

Second, HADPB was not necessary by late 1944 as the Japanese economy was already in tatters as a result of the US naval blockade. Surgical attacks against Japanese steel factories, even if they had been successful, would not have made much of an impact on an economy that was already underproducing by 35 percent.[76]

Third, the emphasis of HADPB on surgical attack against industrial production blinded Hansell to another potential means of paralyzing the

Japanese economy by interdicting its transportation system. Being an island chain, Japan relied upon sea transportation for the movement of intermediate goods. Whereas Germany relied on railways, Japan depended on local shipping. One means by which B-29s could have contributed to the interdiction of Japanese production was through the aerial mining of choke points in Japanese waters, a strategy that did not require good weather or precise high altitude weapons delivery. Yet despite the centrality of shipping to the Japanese economy, Hansell resisted. It would take LeMay to implement the aerial mining campaign, in part because it took relatively little effort and few resources and he had excessive capacity with his expanding B-29 force.[77]

ACTS Contribution to Victory

As with the war against Germany, HADPB did not independently win the war as the ACTS theorists had hoped, though it did significantly contribute to victory. Prewar, HADPB provided the justification to plan for a long-range bomber force. The acquisition process for the B-29, the most technologically sophisticated weapon system of World War II, commenced two years before the United States' entry into the war. Though large numbers of B-29 bombers would not be available until the winter of 1944–1945, without the intellectual and procurement efforts by the Air Corps prior to the war, there would have been no B-29s in 1945 to compel Japan to surrender without an invasion. B-29s may not have been successful in executing HADPB, but the bombers did make a significant contribution to the war's outcome through the aerial mining of Japanese domestic shipping lanes, the firebombing of Japanese cities, and ultimately, the dropping of the atomic bomb.[78]

This chapter highlighted both the limitations and the accomplishments of American strategic bombing in World War II and the critical role the Air Corps Tactical School's unique theory of HADPB played in airmen's thinking leading up to and throughout the war. The predictions of early air power theorists that the advent of the aircraft had changed the very nature of war may have been exaggerated, but the new technology had changed the character of warfare forever. As witnessed in the European and Pacific theaters, victory did not come about by air power alone, but through the combined effects of air, land, and sea forces.

Still, the central tenet of the ACTS theorists that independent air action could decisively win wars has remained a powerful idea for American air power advocates and has diminished little over time, regardless of evidence to the contrary.[79] Powerful, too, has been the assumption that a nation's economic infrastructure is fragile and vulnerable. The air force did not discard its prewar ideas regarding strategic bombing after World War II, a persistence evidenced in the Korean, Vietnam, and Cold Wars. After the Cold War, ACTS strategic bombing theory was expanded to include the targeting of an enemy's leadership and command and control networks, as witnessed in the Instant Thunder air campaign plan of the 1991 Gulf War. Even today, the ACTS focus on vital and vulnerable networks continues to influence air force thinking and doctrine in the air, space, and cyber domains.

Appendix 1
Trenchard Memo

Memorandum from Royal Air Force Chief of Air Staff Hugh Trenchard to Chiefs of Staff Subcommittee on the War Object of an Air Force, May 2, 1928

THE WAR OBJECT OF AN AIR FORCE

At a recent meeting of the Chiefs of Staff Sub-Committee the Report of the Commandant of the Imperial Defence College for the 1st Course (1927) was discussed. In that report the Commandant recommended that the principles of war should be described in identical terms in the Manuals of all three Services. He also expressed the view that at present the situation as regards air warfare was indeterminate. I suggested that the view had arisen from an unwillingness on the part of the other Services to accept the contention of the Air Staff that in future wars air attacks would most certainly be carried out against the vital centres of communication and of the manufacture of munitions of war of every sort no matter where these centres were situated.

It seems to me that the time is now ripe to lay down explicitly the doctrine of the Air Staff as to the object to be pursued by an Air Force in war.

The doctrine which in the past has determined and still determines the object to be pursued by Navies and Armies is laid down in the respective Service Manuals in these words:

(i) *The Navy*: The Military aim of a Navy is to destroy in battle or to neutralize and to weaken the opposing navy including its directing will and morale.

(ii) *The Army*: The ultimate military aim in war is the destruction of the enemy's main forces on the battlefield.

I would state definitely that in the view of the Air Staff the object to be sought by air action will be to paralyse from the very outset the enemy's productive centres of munitions of war of every sort and to stop all communications and transportation.

In the new Royal Air Force War Manual this object will be stated in some such general terms as the following—the actual terms have not been defined: "The aim of the Air Force is to break down the enemy's means of resistance by attacks on objectives selected as most likely to achieve this end."

I will now proceed to examine this object from three viewpoints:

(i) Does this doctrine violate any true principle of war?
(ii) Is an air offensive of this kind contrary either to international law or to the dictates of humanity?
(iii) Is the object sought one which will lead to victory, and in that respect, therefore, a correct employment of air power?

Does This Doctrine Violate Any True Principle of War?

In my view the object of all three Services is the same, to defeat the enemy nation, not merely its army, navy or air force.

For any army to do this, it is almost always necessary as a preliminary step to defeat the enemy's army, which imposes itself as a barrier that must first be broken down.

It is not, however, necessary for an air force, in order to defeat the enemy nation, to defeat its armed forces first. Air power can dispense with that intermediate step, can pass over the enemy navies and armies, and penetrate the air defences and attack direct the centres of production, transportation and communication from which the enemy war effort is maintained.

This does not mean that air fighting will not take place. On the contrary, intense air fighting will be inevitable, but it will not take the form of a series of battles between the opposing air forces to gain supremacy as a first step before the victor proceeds to the attack of other objectives. Nor does it mean that attacks on air bases will not take place. It will from time to time certainly be found advantageous to turn to the attack of an enemy air base, but such attacks will not be the main operation.

For his main operation each belligerent will set out to attack direct those objectives which he considers most vital to the enemy. Each will penetrate the defences of the other to a certain degree.

The stronger side, by developing the more powerful offensive, will provoke in his weaker enemy increasingly insistent calls for the protective employment of aircraft. In this way he will throw the enemy on to the defensive and it will be in this manner that air superiority will be obtained, and not by direct destruction of air forces.

The gaining of air superiority will be incidental to this main direct offensive upon the enemy's vital centres and simultaneous with it.

There is no new principle involved in this attacking direct the enemy nation and its means and power to continue fighting. It is simply that a new method is now available for attaining the old object, the defeat of the enemy nation, and no principle of war is violated by it.

IS AN AIR OFFENSIVE OF THIS KIND CONTRARY TO INTERNATIONAL LAW OR TO THE DICTATES OF HUMANITY?

As regards the question of legality, no authority would contend that it is unlawful to bomb military objectives, wherever situated. There is no written international law as yet upon this subject, but the legality of such operations was admitted by the Commission of Jurists who drew up a code of rules for air warfare at The Hague in 1922–23. Although the code then drawn up has not been officially adopted it is likely to represent the practice which will be regarded as lawful in any future war. Among military objectives must be included the factories in which war material (including aircraft) is made, the depots in which it is stored, the railway termini and docks at which it is loaded or troops entrain or embark, and in general the means of communication and transportation of military *personnel* and *material*. Such objectives may be situated in centres of population in which their destruction from the Air will result in casualties also to the neighbouring civilian population, in the same way as the long-range bombardment of a defended coastal town by a naval force results also in the incidental destruction of civilian life and property. The fact that air attack may have that result is no reason for regarding the bombing as illegitimate provided all reasonable care is taken to confine the scope of the bombing to the military objective. Otherwise a belligerent would be able to secure complete immunity for his war manufactures and depots merely by locating them in a large city, which would, in effect, become *neutral* territory—a position which the opposing belligerent would never accept. What is illegitimate, as being contrary to the dictates of humanity, is the indiscriminate bombing of a city for the sole purpose of terrorising the civilian population. It is an entirely different matter to terrorise munition workers (men and women) into absenting themselves from work or stevedores into abandoning the loading of a ship with munitions through fear of air attack upon the factory or dock concerned. Moral effect is created by the bombing in such circumstances but it is the inevitable result of a lawful operation of war—the bombing of a military objective. The laws of warfare have never prohibited such destruction as is "imperatively demanded by the necessities of war" (Hague Rules, 1907) and the same principle which allows a belligerent to destroy munitions destined to be used against him would justify him also in taking action to interrupt the manufacture and movement of such munitions and thus securing the same end at an earlier stage.

Is This Object One Which Will Lead to Victory, and a Correct Employment of Air Power?

Before I deal with the above heading I would like to state here that, in a war of the first magnitude with civilized nations, I do not for a moment wish to imply by the following remarks that the Air by itself can finish the war. But it will materially assist, and will be one of the many means of exercising pressure on the enemy, in conjunction with sea power and blockade and the defeat of his armies.

In pursuit of this object, air attacks will be directed against any objectives which will contribute effectively towards the destruction of the enemy's means of resistance and the lowering of his determination to fight.

These objectives will be military objectives. Among these will be comprised the enemy's great centres of production of every kind of war material, from battleships to boots, his essential munitions factories, the centres of all his systems of communications and transportation, his docks and shipyards, railway workshops, wireless stations, and postal and telegraph systems.

There is no need to attack the enemy's organised air forces as a preliminary to this direct assault. It will be just as necessary in the future, as it has been in the past, for the Army, assisted by aircraft, to seek out and attack the enemy's Army, but the weight of the air forces will be more effectively delivered against the targets mentioned above rather than against the enemy's armed forces. These objectives are more vulnerable to the attack and generally exact a smaller toll from the attacker.

It will be harder to affect the morale of an Army in the field by air attack than to affect the morale of the Nation by air attacks on its centres of supply and communications as a whole; but to attack—let alone do serious damage—an Air Force in the field is even more difficult. Air bases can be well camouflaged; they can be prepared so that the personnel and material are well protected against bomb attack and their lay-out can be so arranged and spaced as to present a difficult target. An attacker can be induced to waste his strength by deception, such as by dummy aerodromes. Air units can be widely dispersed over the country-side so that it will be difficult to find them and do them extensive damage.

To attack the armed forces is thus to attack the enemy at his strongest point. On the other hand, by attacking the sources from which these armed forces are maintained infinitely more effect is obtained. In the course of a day's attack upon the aerodromes of the enemy perhaps 50 aeroplanes could be destroyed; whereas a modern industrial state will produce 100 in a day—and production will far more than replace any destruction we can hope to do in the forward zone. On the other hand, by attacking the enemy's factories, then output is reduced by a much greater proportion.

In the same way, instead of attacking the rifle and the machine gun in the trench where they can exact the highest price from us for the smallest gain we shall attack direct the factory where these are made.

We shall attack the vital centres of transportation and seriously impede these arms and munitions reaching the battlefield and, therefore, more successfully assist the Army in its direct attack on the enemy's Army. We shall attack the communications without which the national effort cannot be co-ordinated and directed.

These are the points at which the enemy is weakest. The rifleman or the sailor is protected, armed and disciplined, and will stand under fire. The great centres of manufacture, transport and communications cannot be wholly protected. The personnel, again, who man them are not armed and cannot shoot back. They are not disciplined and it cannot be expected of them that they will stick stolidly to their lathes and benches under the recurring threat of air bombardment.

The moral effect of such attacks is very great. Even in the last war ten years ago, before any of the heavier bombers or bombs had really been employed to any extent, the moral effect of such sporadic raids as were then practicable was considerable. With the greater numbers of aircraft, the larger carrying capacity and range, and the heavier bombs available today, the effect would seriously impede the work of the enemy's Navy, Army and Air Forces. Each raid spreads far outside the actual zone of the attack. Once a raid has been experienced false alarms are incessant and a state of panic remains in which work comes to a standstill. Of one town in the last war it is recorded that although attacked only seven times, and that by small formations, no less than 107 alarms were sounded, and work abandoned for the day. Each alarm by day brings day's work to an end—while by night the mere possibility of a raid destroys the chance of sleep for thousands.

These effects, it must be remembered, were produced by occasional raids by very minor forces. The effect on the workers of a Nation of an intensive air campaign will again be infinitely greater than if the main part of that air attack was launched at the enemy's aerodromes and aeroplanes which may be many miles away from the vital points, and, if this air pressure is kept up, it will help to bring about the results that Marshal Foch summed up in the words "The potentialities of aircraft attacks on a large scale are almost incalculable, but it is clear that such attack, owing to its crushing moral effect on a Nation, may impress the public opinion to a point of disarming the Government and thus becoming decisive."

This Form of Warfare Is Inevitable

I have stated above the object which an Air Force should pursue in war, and the reasons on which the Air Staff base their contention that this object is in full accord with the principles of war, is in conformity with the laws of war, and is the best object by which to reach victory.

There is another side to the matter upon which I must lay stress. There can be no question, whatever views we may hold in regard to it, that this form of warfare will be used.

There may be many who, realizing that this new warfare will extend to the whole community the horrors and suffering hitherto confined to the battlefield would urge that the Air offensive should be restricted to the zone of the opposing armed forces. If this restriction were feasible, I should be the last to quarrel with it; but it is not feasible. In a vital struggle all available weapons always have been used and always will be used. All sides made a beginning in the last war, and what has been done will be done.

We ourselves are especially vulnerable to this form of attack; and foreign thinkers on war have already shown beyond all doubt that our enemies will exploit their advantage over us in this respect and will thus force us to conform and to counter their attacks in kind.

Whatever we may wish or hope, and whatever course of action we may decide, whatever be the views held as the legality, or the humanity, or the military wisdom and expediency of such operations, there is not the slightest doubt that in the next war both sides will send their aircraft out without scruple to bomb those objectives which they consider the most suitable.

I would, therefore, urge most strongly that we accept this fact and face it; that we do not bury our heads in the sand like ostriches; but that we train our officers and men, and organize our Services, so that they may be prepared to meet and to counter these inevitable air attacks.

Appendix 2
AWPD-1

Graphic Presentation and a Brief

A-WPD/1 MUNITIONS REQUIREMENTS OF THE ARMY AIR FORCES TO DEFEAT OUR POTENTIAL ENEMIES

THE TASK (Condensed from ABC-1 & Rainbow 5)
1. To wage sustained air offensive against Germany in order to:
   a. Reduce Axis surface and sub-surface operations.
   b. Restrict Axis air operations.
   c. Undermine German combat effectiveness by deprivation of essential supplies, production and communication facilities.
   d. Permit and support a final invasion of Germany.
2. To conduct air operations in strategic defensive in the Orient.
3. To provide air action essential to the security of the continental United States, our possessions and Western Hemisphere.

ACTION NECESSARY TO ACCOMPLISH THE TASK (Air Estimate)
1. Operations against Germany.
   a. Disrupt major portion of electric power, destroy 50 selected targets.
   b. Disrupt German transportation, 47 selected targets.
   c. Destroy 80% of synthetic petroleum establishments, 27 selected targets.
   d. Destroy principal airplane assembly plants, 18 selected targets.
   e. Destroy 90% of sources of aluminum, 6 selected targets.
   f. Destroy major source of magnesium, 6 selected targets.
   (Total selected targets to destroy & keep destroyed) 154.
   g. Provide air support for Army-Navy and joint action incidental to invasion of Germany.
2. Strategic Defensive Operations Against Japan.
   a. Apply air action over northern approach to China seas.

b. Attack or threaten sources of military and civil strength in Japan.
3. Defense of cities' exposed factories, naval and air bases and other critical establishments in the continental United States, each possession and throughout Western Hemisphere against hit-and-run attacks by carriers or flying deckers, against internal disturbances within the Western Hemisphere and action against more remote possibility of the establishment of bases by small hostile forces.

FORCES REQUIRED TO ACCOMPLISH THE TASK IN OPERATIONS AGAINST GERMANY.

(BASIS—Our ability to hit the selected targets will fall off in the same proportion when we move from our peacetime range to actual operations as the RAF Bomber Command experiences when moving from bomb range to actual objectives, day and night, good and bad weather, and opposed by anti-aircraft and pursuit.)

1. To destroy the 154 selected targets during six month operating period 6,834 operating bombardment airplanes are required in accordance with current organization and operating rates. 6,834 operating bombardment airplanes resolve themselves into 98 Groups, 392 Operating Squadrons, 98 Operational Training Squadrons and require 1,708 additional airplanes for Depot Reserve and a monthly replacement of 1,245 airplanes.
2. To protect the air bases occupied by our own striking force (maintaining the same ratio between bombardment and pursuit airplanes that the RAF includes in its ultimate force, 2 Bombers to 1 Pursuit), about 3,400 Pursuit airplanes would be required.

INSUFFICIENCY OF AIR BASES FOR FORCE REQUIRED TO OPERATE AGAINST GERMANY.

1. After the RAF has reached its ultimate strength, there will remain air bases available to units of the Army Air Forces as follows:

Table A2.1.			
Bombardment			
In the U.K.	105 airdromes	42 Groups	(3,026 operational airplanes)
In the Near East	30 airdromes	12 Groups	(816 operational airplanes)
Total Bombardment	135 airdromes	54 Groups	(3,842 operational airplanes)

Table A2.1.

Pursuit			
In the U.K.	25 airdromes	10 Groups	(1,300 operational airplanes)
In the Near East	9 airdromes	6 Groups	(780 operational airplanes)
Total Pursuit	34 airdromes	16 Groups	(2,080 operational airplanes)
Bombardment			
Required	252 airdromes	98 Groups	(6,834 operational airplanes)
Available	135 airdromes	54 Groups	(3,842 operational airplanes)
Shortage	117 airdromes	44 Groups	(2,992 operational airplanes)

2. Since we cannot base in the United Kingdom and the Near East 44 of the total required number of bombardment groups, these 2,992 airplanes required for operations against Germany but for which bases are not available, should be airplanes with a 4,000 mile tactical operating radius of action. Such airplanes are under development, could be employed against Germany from many British or United States controlled points on the Globe converging on Germany from all directions. This force in the field of air warfare would reverse the situation wherein Germany is a continent fighting an island and place Germany in the position of an island in air power under attack from all corners of the earth.
3. The 2,992 required 4,000 mile radius of action airplanes, although under development, cannot be available in necessary quantity earlier than 1945.

EXPEDIENT FORCED BY LACK OF AIR BASE FACILITIES.

1. To initiate an air offensive in 1943 or 1944 (without waiting for the required 4,000 mile radius of action airplanes) double combat crews must be employed. The disadvantages of the use of double combat crews are material but this expediency is unavoidable. As a result, for initial offensive air operations against Germany is required:

Table A2.2.

Aircraft Type	Total Groups	Operating Airplanes	Monthly Replacement
M/B B-25, 26*	10	859	286
H/B B-17, 24	20	1,360	456

Table A2.2.			
Aircraft Type	Total Groups	Operating Airplanes	Monthly Replacement
H/B B-29, * 32	24	1,632	546
Total	54	3,842	1,288

* These airplanes included because of availablility only. Longer range, greated load, carrying airplanes would be far more economical. These airplanes to be replaced by longer range airplanes at the earliest possible date.

2. The "Interim Expedient" Air Force tabulated above would be more than twice as powerful if all bombardment airplanes were B-29 type rather than the type indicated. Again, time requires the use of shorter range, less efficient and much less economical airplanes (per ton of bombs delivered on the objective).

FORCE REQUIRED TO ACCOMPLISH THE TASK OF DEFENDING THE CONTINENTAL UNITED STATES, OUR POSSESSIONS AND THIS HEMISPHERE.

1. <u>Eventual.</u> After the American aviation industry is turning out thousands of bombers monthly, after the Army Air Forces are graduating about 9,000 pilots monthly, and while an offensive is being waged against Germany and Europe, security in the continental United States, each possession and the Western Hemisphere can be assured against any eventuality if the Army Air Forces have available a minimum force consisting of:

Table A2.3. Force Requirements Western Hemisphere				
Location	Bomb Groups	Total Airplanes	Pursuit Groups	Total Airplanes
Continental U.S.	5	(340)	12	(1,560)
(New Eng Area)	(1)	68		
(Norfolk Area)	(1)	68		
(So. Calif.)	(1)	68		
(San Fran Bay)	(1)	68		
Greenland	1	(68)		
Newfoundland	1	(68)	2	(260)
Labrador				

Appendix 2

Table A2.3. Force Requirements Western Hemisphere

Location	Bomb Groups	Total Airplanes	Pursuit* Groups	Total Airplanes
Puerto Rico	1	(68)	1	(130)
British Guiana	1	(68)		
Trinidad			1½	(195)
Saint Lucia				
Antigua				
Brazil	3	(204)		
(Belem)	(1)	68		
(Natal)	(1)	68		
(Rio de Janeiro)	(1)	68		
Other Latin-American and U.S. needs			6	(780)
Chile, Antofagasta	1	(68)		
Peru, Lima	1	(68)		
Mexico, Acapulco	1	(68)		
Panama	2	(136)	2	(260)
Hawaii	3	(204)	3	(260)
Alaska	2	(136)	1½	(195)
Iceland	1	(68)	1	
Bermuda			½	(65)
Bahamas & Jamaica			½	(65)
Philippines	2	(136)	1	(130)
Total	**25**	**1,836**	**32**	**4,160**

* Additional Pursuit units set up among 21 for offensive against Germany to be initially employed in Hemisphere defense as follows:

Table A2.4.		
Air Base Defense, Brazil	3 Groups	390 Aps.
Metropolitan Defense, Rio de Janeiro	1 Group	130 Aps.
Air Base Defense, Chile & Peru	1 Group	130 Aps.
	5 Groups	650 Aps.

SUMMARY
REQUIREMENT OF ARMY AIR FORCES
TO DEFEAT POTENTIAL ENEMY

Table A2.5. Requirement of Army Air Forces to Defeat Potential Enemy

		Force Required					
		Interim Expedient Force			Ultimate Force		
Air Force Missions	Type Aps	Groups	Airplanes (Including Depot Res)	Monthly Replacement in Aps	Groups	Airplanes (Including Depot Res)	Monthly Replacement in Aps
Air Offensive against Germany	B-25 & 26	10	1,062	286	10	1,062	143
	B-17 & 24	20	1,700	456	20	1,700	228
	B-29	24	2,040	546	24	2,040	273
	4,000 mi.				44	3,740	501
	Day Pur.	21	2,756	334	21	2,756	334
	Night Pur.		656	80		656	80
Defend U.S., Possessions & Hemisphere	B-17 & 24	33*	2,805	70	25	2,125	81
	Day Pur.	32	4,200	114	32	4,200	90
	Night Pur.		1,000	27		1,000	27

Table A2.5. Requirement of Army Air Forces to Defeat Potential Enemy

Air Force Missions	Type Aps	Force Required					
		Interim Expedient Force			Ultimate Force		
		Groups	Airplanes (Including Depot Res)	Monthly Replacement in Aps	Groups	Airplanes (Including Depot Res)	Monthly Replacement in Aps
Strategic Defensive in Asia	B-17 & 24	2	170	21	2	170	34
	Day Pur.	1	132	17	1	132	17
	Night Pur.		31	4		31	4
Air Support for Ground Forces (Appropriate to Size Ground Force Set Up in Tab D)	A-20 D/	13	946	42	13	946	42
	Bombers	13	1,255	56	13	1,255	56
	Obsn.	108 SQ	1,901	98	108 SQ	1,901	98
	Photo	2	142	23	2	142	23
	Transport	19	1,520	77	19	1,520	77
	Gliders	**	(3,000)	(153)		(3,000)	(153)
Aaf Maintenance	Transport	13	1,040	25	13	1,040	25
Training	All Types	—	37,051	—	—	37,051	—
Grand Total		203* 108 SQ	59,727	2,276	239 108 SQ	63,467	2,133

* (Including 8 Groups [680 airplanes] not included in total)
** (Not included in totals)

Table A2.6. Interim and Ultimate Force			
Interim Expedient Force		**Ultimate Force**	
90,391 Pilots AC		103,482 Pilots AC	
31,186 Combat Crew AC		32,044 Combat Crew AC	
39,214 Non-Flying AC		40,798 Non-Flying AC	
18,607 Other Branch	179,398	**19,355** Other Branch	195,679
813,951 AC-Technicians		862,439 AC Technicians	
588,267 AC Non-Technical		614,413 AC Non-Technical	
479,503 Other Branch	1,881,721	**492,395** Other Branch	1,969,247
	2,061,119		2,164,916

2. <u>Immediate</u>. A large measure of security assured under the conditions stated above is due to the output rate of airplanes and trained personnel stated plus the fact that a period of about six months would be required by the Germans after winning the war in Europe and before changing to an offensive against the Western Hemisphere. <u>Until those output rates are established</u>, even though an offensive is being waged against Germany, security against any eventuality cannot be guaranteed without an additional force of at least 8 Heavy Bombardment Groups (544 total airplanes less reserve) in the Western Hemisphere. These 8 Groups would constitute a reserve to meet a threat greater than that conveyed by 1 or 2 ships or to "bomb them out" in the event that a small force established a base in the Western Hemisphere. This 8 group general reserve would probably be located in the United States the greater portion of the time.
3. <u>General</u>. If no offensive is being waged against Germany in Europe, a much larger force than that set up in the foregoing will be essential if the security of the continental United States, each of our possessions and the Western Hemisphere is to be assured against any eventuality.

<u>FORCE REQUIRED TO ACCOMPLISH THE TASK OF STRATEGIC DEFENSIVE AIR OPERATIONS IN ASIA.</u>

1. <u>Conception</u>. If a shortage of airplanes were not the controlling factor, a satisfactory strategic defensive against Japan could be most effectively accomplished by a vigorous air offensive against the sources of Japan's military and civil strength. Actualities force the modification of that con-

ception to the creation of a striking force designed to deter Japanese expansion and aggression in Asia.

2. <u>Minimum immediate requirements</u>. In addition to the forces set up for the defense of the Philippines and Alaska, it will be necessary to add 2 Heavy Bombardment Groups (136 total airplanes less reserve) and 1 Pursuit Group (130 airplanes less reserve) to the air garrison of the Philippine Department to attack or to threaten Japanese sea communications in the China Seas and Japanese source of military and civil strength.

FORCES REQUIRED TO ACCOMPLISH THE TASK OF AIR SUPPORT TO THE ARMY.

1. <u>Basis</u>: An invasion by the United States Army of Germany or German controlled areas will not be attempted until the air offensive against Germany has been successful and we can control the air over the area of invasion to a very large measure. By that time, a large proportion of the force set up for the air offensive against Germany will be available to support invading surface forces. Hence, the force set up below is the bare minimum to permit the training of the invading forces and to provide limited support in violent ground actions such as the suppression of disturbances in the Hemisphere or the ejecting of small hostile forces from the Hemisphere or our possessions.

2. The Air Support Force appropriate to a ground Army of the size set up in this paper is as follows:

Table A2.7. Air Support Force				
Location	L/B	D/B	Trans.	Obsn.
1st Air Support Command	2 Grps	2 Grps	13 Grps	5 Sqs for
(First Army)			(832 Aps)	Armies,
2nd Air Support Command	2 Grps	2 Grps		54 Sqs for
(Second Army)				Corps,
3rd Air Support Command	2 Grps	2 Grps		39 Sqs for
(Third Army)				Armored Div.
4th Air Support Command	2 Grps	2 Grps		
(Fourth Army)				
5th Air Support Command	2 Grps	2 Grps		
(Armored Force)				

Table A2.7. Air Support Force			
Location	L/B	D/B	Trans.	Obsn.
Iceland	1 Sq	1 Sq		1 Sq
Brazil	1 Gp	1 Gp		4 Gps
Puerto Rico	1 Sq	1 Sq		1 Sq
Trinidad	1 Sq	1 Sq		
Panama	1 Sq	1 Sq		1 Sq
Columbia-Ecuador-Peru	1 Sq	1 Sq		1 Sq
Alaska	2 Sq	2 Sq		1 Sq
Hawaii	1 Sq	1 Sq		1 Sq
Philippines	2 Sq	2 Sq		1 Sq
Caribbean Defense Command			1 Gp	
Continental United States & Strategic Reserves			13 Gps	
GHQ			2 Gps Photograph (mapping)	
3,000 (600 of which are Depot Reserve) 15 Man Gliders, distribution undetermined			
1 Pursuit Group per each Support Command is included among the 6 Pursuit Groups set up for other Latin American and U.S. defense needs.			

FORCE REQUIRED TO PROVIDE AIR SUPPORT FOR THE AIR FORCES.

In order to maintain the services required by the air forces set up in the foregoing using the minimum amount of air transports estimated as essential, there are required 13 Transport Groups (1,040 transports, including Depot Reserve to consist of 160 4-engine long range and 880 2-engine medium range transports, all with approximately 30-man capacity).

TRAINING REQUIREMENTS TO BUILD AND TO MAINTAIN THE ARMY AIR FORCES.

1. The critical element insofar as training of personnel is concerned, consists of the training of pilots requiring to form and to maintain the Air Forces set up above. While the Interim Expedient Air Force is operating, it is estimated that attrition will require the graduation of 85,236 pilots per year. In order to expand our pilot training establishment from the pres-

ent objective (30,000 graduates per year) to an 85,236 rate, 57 new flying schools will be required.

2. When the ultimate Air Force is organized and the necessity for operations by double combat crews no longer exists, the annual replacement requirements are estimated as 108,528. This rate will require 72 additional flying schools beyond those set up to graduate 30,000 a year.

3. Flying schools to graduate 85,236 per year would be set up immediately upon the approval of this program. These schools would be expanded to the 108,528 rate approximately one year before the ultimate Air Force could be established. It will be noted that this date is indefinite but considerably earlier than the ultimate Air Force would set up under that heading in war as far as the procurement of airplanes is concerned.

REQUIREMENTS IN TRAINING AIRPLANES

Training airplanes by number and type will be required to produce the numbers of graduated pilots annually, in accordance with the table below:

	Table A2.8. Operating Training Type Airplanes	
Type	85,236 Graduated Pilots per Year for Interim Expedient Air Force	108,528 Graduated Pilots per Year for Ultimate Air Force
Primary Trainer	8,520	11,115
Basic Trainer	8,804	11,486
Adv. Trainer-1 Engine	2,954	3,853
Adv. Trainer-2 Engine	5,736	7,484
Bombardment	1,988	2,594
Observation	341	445
Transport	57	74
Total Training	28,400	37,051

TIME TO MEET TRAINING REQUIREMENTS.

As a general statement, pilots can be trained prior to the dates that the combat airplanes which the pilots are to operate can be manufactured.

For example, if the necessary numbers of additional Primary Training type airplanes (4,195) can be added to those expected to be on hand (4,325 plus 4,195 equals 8520) on February 15, 1942, it would be possible to start the expanded pilot training schools in operation on May 15, 1942. It would then be possible to man and to meet war time attrition on the Interim Expedient Air Force by November 1, 1943, without interfering with the 54 Group Program. November 1, 1943, however, is far ahead of the date on which it is possible to manufacture all of the combat airplanes included in the Interim Expedient Air Force.

PERSONNEL REQUIREMENTS.

To man and to operate Army Air Forces of the size and composition set forth in the foregoing, to operate all training, maintenance and overhead installations necessary therewith, there will be required personnel approximately as follows:

Table A2.9. Personnel Requirements				
Officers		Interim Expedient		Eventual
Pilots		103,482		90,391
Navigators		21,462		19,866
Bombardiers		7,387		8,125
Observers		3,195		3,195
	Total Flying	135,526		121,577
Non-Flying AC		40,798		39,214
	Total AC	176,324		160,791
Other Branches		19,355		18,607
	Total	**195,679**		**179,398**
Enlisted Men				
Technicians AC*		862,439		813,951
Non-Technical AC		614,403		588,267
	Total AC	1,476,842		1,402,218
Other Branches		492,395		479,503
	Total	**1,969,237**		**1,881,721**
Grand Total		**2,164,916**		**2,061,119**
* Approximately one year required for training to degree necessary for operations. Twenty-two new technical schools will be required therefor. These numbers include approximately 50,000 enlisted members of combat crews.				

MANUFACTURE OF AIRPLANES.

The numbers of additional factories and the output of airplane factories is determined by the monthly replacement rates required to sustain the Air Forces set forth in the foregoing in war operations. The total number of airplanes of each type which must be built (in addition to current orders) to build and to maintain the Army Air Forces in one year of operations are included in table A2.10 to the compilation by the Joint Board.

Table A2.10. Monthly Aircraft Replacement		
Aircraft Types	Interim Expedient Monthly Replacement Required	Ultimate Requirement Monthly Replacement Required
M/Bomb B-26	286	143
H-Bomb B-17	547	319
H/Bomb B-29	546	273
H/Bomb 4,000 mile		501
Dive & Light	98	98
Total Bombardment	**1,477**	**1,334**
Pursuit	576	576
Observation & Photo	121	121
Transport	102	102
Grand Total	**2,276**	**2,133**

POSSIBILITY TO MANUFACTURE WITHIN REASONABLE TIME.

It is within the capacity of the nation to implement the Army Air Forces indicated in the foregoing.

An air offensive against Germany can be initiated by a token force of 3 Heavy Bombardment Groups in April 1942.

If an all out effort is initiated immediately, the air offensive against Germany can reach full power in April 1944.

The air offensive against Germany can be sustained at full power.

COMPARISON OF OPPOSING AIR FORCES.

In operations against Germany, the estimated numerical relationship between the Germany Air Force and the combined strength of the Army Air Forces and the Royal Air Force will be as follows:

Table A2.11. Comparison of Opposing Air Forces		
	Superiority GAF	RAF + AAF
July 1, 1943		
Bombers	62%	
Pursuit	22%	
Fighter	90%	
Total Combat	83%	
Mid 1944		
Bombers		15%
Pursuit	22%	
Fighter	90%	
Total Combat	16%	
Ultimate		
Bombers		59%
Pursuit	22%	
Fighter	90%	
Total Combat		11%

The numerical comparison is no true indication of the relative strength of the opposing Air Forces. We now have a superiority in <u>performance</u> in Heavy Bombers and it is proposed to maintain that superiority and win the war by exploiting the superiority of our Heavy Bombers to the utmost.

The grand total Army Air Forces set up in the foregoing is a large force. It is designed to win a large war.

CONCLUSION.

The forces required to accomplish the mission of the Army Air Forces have been computed carefully. They represent the minimum force necessary to succeed in each task. Any reduction in force must therefore include a reduction in task.

For example, if the 25 groups required for the defense of the United States, our possessions and the Western Hemisphere cannot be made available, the task must be reduced to the defense of a specific part of the Hemisphere or of specified possessions.

If the required combat airplanes are manufactured for the Army Air Forces, personnel can be trained to man and to operate those air forces as soon as the combat airplanes can be manufactured.

The Army Air Forces are prepared to accomplish their mission in the defeat of our potential enemies as soon as American aviation industry can manufacture the required airplanes.

AIR OPERATIONS OTHER THAN THOSE DIRECTED AGAINST GERMANY PROPER

(Army Air Force munitions requirement to defeat our potential enemies)

STRATEGIC OBJECTIVE: To defeat Germany (and her Allies).

ARMY AIR FORCE TASKS: (In addition to operations against Germany proper)

a. To provide air action essential to the security of the continental United States, our possessions and the Western Hemisphere.
b. To combat air operations in strategic defensive in the Orient.
c. To provide close support to the ground forces during training, defensive operations and eventual invasion.

FORCES REQUIRED TO ACCOMPLISH THE TASKS:

a. Security. The forces provide the minimum coverage required to—
  (1) Protect vital economic and military objectives against carrier, flying deck or land based air attack;
  (2) Defend the Pacific Triangle, Alaska and approaches to the Panama Canal;
  (3) Acquire data and information essential for all out defense in the event of a forced retirement from the European Theatre.

The forces set up are minimum. Note that the forces include only 5 Bombardment Groups in the continental United States and that coverage in South America is very thin.

Pursuit forces are set up to provide limited defense over the areas indicated and to defend bombardment bases.

Total Force Required:

Bombardment	25 Groups 1,700 Airplanes (Not including Depot Reserve)
Pursuit	32 Groups 4,160 Airplanes (Not including Depot Reserve)

In addition to these units, 8 Groups of Heavy Bombardment must be available as general reserve until maximum output of airplanes and pilots is underway.

b. Strategic Defensive in Orient. The forces set up solely for the defense of the Philippines include 2 heavy bombardment groups, are to be augmented by 2 additional heavy bombardment groups and 1 pursuit group. This augmented force can operate against the northern approaches to the China Sea and from the bases indicated in the map should go far toward discouraging the Japanese from any offensive operations in the Orient.

c. Close Support. The support forces set up are the requirement of the ground force establishment set forth in table A2.12 to the Joint Board estimate.

(1) Total Support Force Required:

Table A2.12. Total Close Support Requirements		
Light Bombers	13 Groups	757* airplanes
Dive Bombers	13 Groups	1,004 airplanes
Pursuit	5 Groups	650 airplanes
Observation	108 Sqs	1,521 airplanes
Transport	19 Groups	1,216 airplanes
Gliders	15 Man	2,400 airplanes for 10 Air Force
* (Not including Depot Reserve)		Divisions & 64 Parachute Batallions

AIR OFFENSIVE AGAINST GERMANY

(Army Air Forces' Munitions Requirements to Defeat our Potential Enemies.)

STRATEGIC OBJECTIVE: To defeat Germany (and her Allies).

THE ARMY AIR FORCE TASK.

TO: a. Destroy the industrial war making capacity of Germany.
         b. Restrict Axis air operations.
         c. Permit and support a final invasion of Germany.

Table A2.13. Action Necessary to Accomplish the Task	
No. Selected Targets	Results
50	Disrupt electric power
47	Disrupt transportation
27	Destroy 80% synthetic petroleum
18	Destroy airplane assembly plants
6	Destroy 90% aluminum
6	Destroy magnesium
154 Total targets to destroy and keep destroyed to accomplish the task	

FORCE REQUIRED TO ACCOMPLISH THE TASK

The exact number of airplanes required to assure the complete destruction of these 154 selected targets has been determined by a detailed study of bombing accuracy in wartime operations including pursuit and antiaircraft opposition. This approach and analysis has established the requirement that 6,834 operating bombardment airplanes are required to accomplish the task during the six month period that weather conditions favor operations over Germany.

However, absence of necessary bases and time required to design and manufacture necessary number of 4,000 mi. radius of action bombers combine to force use of double combat crews. These conditions are met by an Interim Expedient Force consisting of the following:

Table A2.14. Interim Expedient Force					
Based in United Kingdom					
Units	Operating Airplanes		Monthly Replacement in Airplanes		
10 Groups (B-25 & 26*)	850 Bomb		286 Bomb		
20 Groups (B-17 & 24)	1,360 Bomb		456 Bomb		
12 Groups (B-29 & 32)	816 Bomb		273 Bomb		
10 Groups Pursuit		Pur. 1,300		Pur. 209	
Based in Near East					
12 Groups (B-29 & 32)	816 Bomb		273 Bomb		
6 Groups Pursuit	———	Pur. 780	———	Pur. 126	
Total	3,842 Bomb	2,080 Pur.	1,288 Bomb	335 Pur.	
* These airplanes included because of availability only. Longer range, greater load carrying airplanes would be far more economical. These airplanes to be replaced by longer range airplanes at the earliest possible date.					

November 18, 1941
MEMORANDUM FOR THE CHIEF OF THE AIR STAFF

Subject: Notes on the Preparation of AWPD/1

1. The latter part of July, Colonel Bissell of the WPDS brought into General Arnold a copy of a letter from the President addressed to the Secretary of War, requesting the preparation of the Over-All Production requirements required to defeat our potential enemies. Colonel Bissell suggested that the air part of the problem could be approached in one of the following ways: by the War Plan and Air War Plans Division making independent studies and then coordinating them; by the War Plans Division with the assistance of the AWPD preparing the plan. This latter method was approved by General Arnold.
2. Colonel Bissell then contacted Colonel George and preliminary meetings were held involving the latter two officers, Major Kuter, Hansell, and Lt.

Colonel Walker. During these initial conferences the procedure desired by the Air War Plans Division in approaching the project was put into effect.

3. The AWPD, recognizing its responsibilities under AR 95-5, determined that the matter was primarily an air matter and should be prepared by AWPD as a staff agency of the Chief of the Air Force rather than as subordinate assistant to a section of the War Plans Division General Staff.

4. Consequently, from that time on the entire project was handled exclusively under the direction of the AWPD. Colonel Bissell attended a few meetings but actually the War Plans Division of General Staff thru him did not contribute to the preparation of this plan.

5. A memorandum was prepared by AWPD on August 4, wherein the method of procedures was outlined and the responsibilities delegated to representatives of the A-2, A-3, A-4, T & O Division OCAC, Materiel Division OCAC. Each of the several agencies were requested to prepare the pertinent information essential to the plan.

6. All data which formed the basis of AWPD/1 was secured almost exclusively thru Army Air Forces agencies. It is noteworthy that the entire Intelligence background was gained from the Intelligence Division OCAC which had for the preceding year been assembling and preparing that character of data essential to such a plan as AWPD/1.

7. The entire plan was predicated upon the best determination possible of the mission and the means required with which to accomplish it. Improving this information the calculation of equipment was calculated.

8. After determination of the aircraft was reached it was then necessary to tabulate all related aviation material. A representative of the A-4 Division proceeded to Dayton with officers called in from there for instruction and this latter information was compiled at Wright Field.

9. On August 11, the study was completed and on that date was presented orally in brief to the Assistant Chief of Staff, G-3, and to principal assistants who indicated general concurrence. The Chief, Army Air Force, was informed that the study was ready for his consideration.

10. On August 13, the study was presented orally to Mr. Lovett, General Arnold, Assistant Chief of the Air Staff WPD, and the Chief of Air Staff.

11. On August 22, the study was again presented to General Brett, General Fairchild, and Colonels Bundy, Wilson and Chauncey of Plans Division. On August 30, study was presented to General Marshall, Mr. W. A. Harriman, General Arnold, General Fairchild, and Colonel Bundy. Within an

hour after the presentation of the study, the AWPD was informed that General Marshall had approved the study and had written "O.K. GOM" under General Arnold's signature on the original copy.

12. While it was intended to present this study thru the War Plans Division of the General Staff which had been provided copies of the study thru Colonel Bissell, it was presented directly by General Arnold to General Marshall. On September 4th, the study was presented to the principal officers of the OPM including Messrs. Knudson, Harrison, Batt, Johnston, Heigs, Britton, and Bigger. These officials were accompanied by Generals Marshall and Arnold. On September 11, Mr. Stimson called Colonels George and Walker and Major Kuter into his office for about an hour and a half of information discussion of the study. He implied his approval as a practicable project only if the nation were at war. On September 12, the study was presented to Mr. Stimson and Mr. McCloy. Following this presentation, the study was again presented to 27 key members of the air Staff, OCAC, and Air Force Combat Command.

13. On September 15, the study was presented to Colonel W. M. Donovan, Chief of Intelligence for the President and General Spaatz. Since that date, until the final compilation by the WPD General Staff "Army and Navy Estimate of U.S. Over-All Production Requirements" representatives of the AWPD conferred with Colonel Scobey of the Joint Board in preparation of the above.

Appendix 3
AWPD-42

THE WHITE HOUSE
WASHINGTON

August 24, 1942

SECRET

MEMORANDUM FOR GENERAL MARSHALL:

I wish you would ask General Arnold to submit to you his judgment of the number of combat aircraft by types which should be produced for the Army and our Allies in this country in 1943 in order to have complete air ascendency over the enemy.

This report should be prepared without consideration for existing schedules or production possibilities or any other competing military requirements. I am asking for this because I would like to know what the theoretical requirements are to get complete control and domination of the air.

I realize fully, however, that there are limiting factors to the creation of air power, such as the availability of pilots, high octane gas, transportation and the competition of other essential critical munitions of war. Hence, I would like you and Admiral King to submit a second schedule based on these realities and the proper relationship of air power to the Navy and our ground forces.

/s/FRANKLIN D. ROOSEVELT

September 9, 1942

MEMORANDUM FOR THE CHIEF OF STAFF:

Subject: Combat Aircraft Which Should be Produced in the United States in 1943.

1. Pursuant to the instruction from the President in his memorandum for you, dated August 24, 1942, an estimate has been made of the number of combat aircraft by type which should be produced, for the Army, the Navy, and our Allies, in this country in 1943 in order to secure complete air ascendancy over the enemy.
2. The requirements have been based upon the following air operation in 1943 and early 1944:
    a. An air offensive against Europe to deplete the German Air Force, destroy the sources of German submarine construction and undermine the German war-making capacity.
    b. Air support of a land offensive in Northwest Africa.
    c. Air Support of United Nations land operations to retain the Middle East.
    d. Air support of surface operations in the Japanese Theater to regain base areas for a final offensive against Japan Proper, including:
        (1) Land operations from India through China, reopening the Burma Road.
        (2) Amphibious operations from the South and Southwest Pacific toward the Philippine Islands.
    e. Hemisphere Defense, including anti-submarine patrol.
3. To implement these air operations, the following airplanes should be produced in the United States in 1943:

Table A3.1.				
	U.S. Army	U.S. Navy	Others	Total
Tactical	63,068	24,800	19,540	107,408
Training	12,232	8,000	1,900	22,132
Liaison	116	250	1,000	1,366
Total Airplanes	75,416	33,050	22,440	130,906
Gliders 8,284				

4. These air operations require the development and deployment of the following Army Air Forces by January 1, 1944:

Table A3.2.				
Aircraft Type	Groups	Airplanes	Gliders	Air Transport Command, long range transports
Heavy Bomb	76	3,848		
Med Bomb	43	2,752		
Light Bomb	14	896		
Dive Bomb	12	1,152		
Ftr.	70	7,000		
Obs.	20	1,680		
Photo Recon.	12			
Troop Carrier	74	624		
Glider		1,768	8,284	
Total	281	19,520	8,284	2,217

5. Personnel requirements—Strength of the A.A.F. by January, 1944.

Table A3.3.			
	Army Air Corps	Other Branches	Total
Officers	230,243	72,600	302,843
Enlisted Men	1,554,104	877,400	2,431,504
Total	1,784,347	950,000	2,734,347

6. Logistical requirements, Army Air Forces.
    Bombs   1,140,363 tons
    Gasoline 4,888,941,000 gallons
    Shipping 17,421,507 ship tons, total during 1943.

7. Details concerning these requirements are contained in the body of the report and in the Annexes.

//signed//
H. H. ARNOLD,
Lieut. General, U.S.A.
Commanding General, Army Air Forces.

PART IV
REPORT
1. Directive.

Determine the number of combat aircraft by type which should be produced in this country in 1943 in order to have complete air ascendency over the enemy (extract from a letter from the President to General Marshall, August 24, 1942).

2. Definition.

Air ascendency: the conditions OF air strength, both of ourselves and of the enemy, under which it will be possible for our several armed forces to complete the defeat of our enemies.

Under this definition it will be observed that: (1) the enemy air strength must be so depleted as to render him incapable of frustrating the operations of our air, land, and sea forces; and (2) our own air strength must be so developed as to permit us to carry out the roles of our air force, in conjunction with our land and sea forces and also independently thereof, which are necessary for the defeat of our enemies.

3. Strategic situation and concept.

European

By the time that the air forces contemplated in this study are ready for employment, it is likely that large Axis ground forces will be released from the Russian front for employment elsewhere. Under these circumstances the ground forces of the United Nations will be numerically inferior to the Axis ground forces in Europe. If our ground forces, which are numerically inferior, are to defeat the seasoned troops of the Axis in Europe, then circumstances must be created which will make this possible. Our numerically superior air forces must deplete the air forces of the enemy and undermine the structure which supports his surface forces. Fortunately a base, England, is available to us which is capable of sustaining our increasingly superior air power, and is within striking distance of the sources of German air power and the vitals of the German war economy.

Far Eastern

Our armed forces in the Far Eastern theater are not within effective striking distance of the vital sources of Japanese military policy. Unless the Russian Maritime Provinces can be made available—and retained—as bases of operation, we will be unable initially to wage a sustained air offensive against Japan. This condition cannot be relied upon. Hence our land and sea forces, supported by our air forces, must recover lost areas which are suitable as offensive bases against Japan proper. When these bases have been recovered, then our air power can be brought to bear against the highly vulnerable structure of Japan. Hence from the standpoint of air requirements, the Far Eastern operations may be divided into two phases:

(1) Air operations in support of our land and sea forces to regain bases within striking distance of Japan. This involves support of amphibious forces driving northwest from Australia as a base area, and of land forces driving northeast from India as a base area.

(2) Air operations against Japan proper to destroy her war making capacity. This operation may be undertaken fairly promptly if it is possible to retain the Russian Maritime provinces as a base area.

Sequence of Operations in 1943 and early 1944.

Air Operations:

Since:

(1) The German air force must be depleted and the German war economy must be undermined before a successful invasion of the European continent can be undertaken; and

(2) Base areas for an air offensive must be secured before a decisive attack can be launched against Japan; and

(3) Each of these undertakings will involve large forces and will require considerable time for accomplishment.

It appears that the air operations which can be carried out simultaneously in 1943 and early 1944 may be listed as follows:

Program A

(1) An Air offensive against Germany to deplete the German Air Forces and submarine force and undermine German war economy.

(2) Air support of operations in North Africa.

(3) Air support of operations in Middle East.

(4) Air operations in Far East. Support of surface forces in regaining bases and operations against enemy lines of communication and installations from available bases.

(5) Air operations in Hemisphere Defense.

When these operations have been successfully accomplished, we will be in a position to carry out the following air operations—later in 1944—successively and simultaneously.

Program B

(6) Air operations in support of a Combined Offensive against Germany.
(7) An air offensive against Japan.

4. <u>Description of air operations.</u>

   a. <u>AIR OFFENSIVE AGAINST GERMANY.</u>

The air offensive against Germany is a combined effort by the U.S. Army Air Force and the R.A.F. The U.S. Army Air Force will concentrate its efforts upon the systematic destruction of selected vital elements of the German military and industrial machine through precision bombing in daylight. The R.A.F. will concentrate upon mass air attacks of industrial <u>areas</u> at night, to break down morale. In view of the acute shortage of skilled labor in Germany this effort of the R.A.F. should have a pronounced effect upon production.

Systems of objectives to be destroyed and priorities are as follows:
First Priority: Destruction of the German Air Force.
    Targets: 11 fighter factories; 15 bomber factories; 17 airplane engine plants.
    Destruction: Complete—with repeated attacks at two month intervals.
    Results: Almost complete destruction of the sources of German air power, with consequent depletion of the German air force through combat attrition caused by these—and other—bombing raids.
    Bomber force: 22,374 bomber sorties.
    Bombs: 44,748 tons (100 times the tonnage dropped on Renault)
Second Priority: Submarine building yards.
    Targets: 20 building yards.
    Destruction: Complete—one attack each.
    Results: Germany's submarine shipbuilding program completely disrupted. This <u>offensive</u> cure to the submarine menace, at its source, is the only conclusive solution. Other types of antisubmarine operations are defensive and inconclusive.
    Bomber force: 10,332 bomber sorties.

Bombs: 20,664 tons.

Third Priority: Transportation

Targets: 38 (locomotive building shops; locomotive repair shops; marshaling yards; inland waterways).

Destruction: Partial.

Results: Breakdown of a vital link in the German military and industrial structure—one which is at present taxed to its maximum capacity and has become very sensitive to disruption.

Bomber force: 9,348 sorties.

Bombs: 18,696 tons.

Fourth Priority: Electric Power

Targets: 37 major electric power plants.

Destruction: Of targets selected—complete.

Results: Virtual paralysis of the major manufacturing centers. Germany is now working her extensive power system to the limit. Loss of such a tremendous source of energy would have immediate and wide-spread effect. However, harassing raids must be repeated in order to keep these areas isolated from other sources of electric power.

Force Required: 13,447 bomber sorties.

Bombs: 26,894 tons.

Fifth Priority: Oil.

Targets: 23 plants.

Destruction: Complete.

Results: Reduction of 47% of Germany's refined oil products.

Force required: 8,322 bomber sorties.

Bombs: 16,644 tons.

Sixth Priority: Alumina.

Targets: 14 plants.

Destruction: Complete.

Results: Loss of practically all aluminum production in Germany and occupied countries. This would be a severe blow, since aluminum is now extensively used as a replacement for copper, of which there is an acute shortage.

Force required: 1,932 bomber sorties.

Bombs: 3,864 tons.

Seventh Priority: Rubber.
>Targets: 2 synthetic (Buna) plants.
>Destruction: Complete.
>Results: The loss of approximately 48% of rubber supply, to Germany. Immediate effect upon all forms of the armed services.
>Force required: 288 bomber sorties.
>Bombs: 576 tons.

Recapitulation:
>Targets: 177.
>Force required: 66,045 bomber sorties.
>Bombs: 132,090 tons of bombs.
>Results: Decimation of the German Air Force.
>>Depletion of the German Submarine Force.
>>Disruption of German war economy.

b. AIR SUPPORT OF OPERATIONS IN NORTH AFRICA, with partial opening of the Mediterranean and a base of operations against Italy.

c. AIR SUPPORT OF OPERATIONS IN THE MIDDLE EAST, to hold the Middle East and drive the Axis forces out of Africa.

d. AIR OPERATIONS IN THE JAPANESE THEATER. Support of a land offensive to reopen the Burma road and gain operating bases in China.
>Support of an amphibious offensive to regain the Philippines.
>Support of land forces holding Siberia, if possible.

e. AIR OPERATIONS IN HEMISPHERE DEFENSE. Primarily the defense of the American Republics against carrier attacks, and the defense of shipping by air operations against submarines.

f. AIR OPERATIONS IN SUPPORT OF A COMBINED OFFENSIVE AGAINST GERMANY. This involves the provision of additional fighters, light and dive bombers, observation, and transports for the close support of a land invasion of Europe from the British Isles. This operation must be subsequent to a successful air offensive.

g. AN AIR OFFENSIVE AGAINST JAPAN. Considering the great distances involved, it is apparent that the majority of our bombing effort must be carried out by long-range bombers (B-29 type). Those will not be available in quantity until late in 1944. The following table indicates the system of targets selected and the effect of destruction of each. The total force required for this offensive is 51,480 bomber sorties.

Table A3.4. Targets

Appendix	System of Targets	Number of Targets	Percentage of total Production Represented by Targets
J I	Aircraft and Engine	14	78.1
J II	Submarine Yards	5	100
J III	Naval and Commercial Bases	20	99.2 (Naval); 92.7 (Commercial)
J IV	Alumina and aluminum	20	100 (Alumina); 77.1 (Aluminum)
J V	Iron and Steel	21	100 (Iron); 94.3 (Steel)
J VI	Oil	15	87
J VII	Chemicals	14	—
J VII	Rubber	14	100
	Total Number of Targets	123	

5. <u>Factors Involved in conducting those air operations.</u>
   a. Destructive effect of bombing. Direct hits by bombs will destroy all the targets selected. In some cases repeat-raids must be conducted to prevent rebuilding. Forces have been provided to meet this requirement.
   b. Feasibility of conducting accurate bombing. Experience has shown that it is perfectly feasible to conduct accurate, high level, daylight bombing under combat conditions, in the face of enemy antiaircraft and fighter opposition.
   c. Feasibility of penetrating fighter and AA defense without excessive losses. With our present types of well armed and armored bombers, and through skillful employment of great masses, it is possible to penetrate the known and projected defenses of Europe and the Far East without reaching a loss-rate which would prevent our waging a sustained offensive.
   d. Rate of operations, and weather. Studies of the European and Japanese Theaters indicate that the following rates of operation of bomb-

er units may be anticipated: Europe—5 to 6 operations a month. Far East—10 operations per month.

6. <u>Air Forces Required to carry out the operations listed above in 1943 and early 1944.</u>

281 Groups, to carry out operations 1 to 5 incl., Program A. The size force required to fully complete this task cannot be provided in the theaters shown until January 1944. However, during the period of build-up—in 1942—the available forces can be partially completing the selected operations. It is anticipated that they can complete about one-third of the tasks required for the air offensive in 1943. Hence, it is expected that the air offensive against Germany, requiring six months of operations of the complete force, at the rates of operations expected, can be one-third accomplished in 1943, thus requiring four months of operations in 1944. This operation should be complete by May 1944, and the Combined Offensive should follow immediately thereafter.

336 Groups, to carry out operations 1 to 7 incl., Program B.

7. <u>Recapitulation of Combat Aircraft Required.</u>

To carry out operations 1 to 5 incl., (Program "A"), the U.S. Army Air Forces will require 63,068 tactical aircraft in 1943.

To carry out operations 1 to 7 incl., (Program "B"), the U.S. Army Air Forces will require 74,944 tactical type aircraft in 1943.

8. <u>Air Bases.</u>

There will be ample air bases in the United Kingdom to accommodate the air forces set up for the European Theater.

In the Japanese Theater there are at present insufficient air bases to accommodate the land based air forces which are deployed in this study. It will be necessary to construct:

24 new bases in the Central Pacific.
20 new bases in the South Pacific.

It is believed that the deployment shown in this study represents virtual saturation of the Japanese Theater, and that larger air forces cannot be accommodated without an extensive air-base building program.

9. Total Aircraft Required, including trainers and replacements, for the U.S. Army Air Forces in 1943.

Table A3.5. Total Aircraft Required

Program A			Program B
63,068	Tactical airplanes	74,944	Tactical airplanes
12,232	Training airplanes	22,716	Training airplanes
8,284	Gliders	10,499	Gliders
116	Liaison planes	828	Liaison planes
83,700	**Total**	**108,987**	**Total**

In accordance with the established policy in such matters, it is anticipated that the requirements of the U.S. Army Air Forces for Army Type aircraft will be given first priority in allocation of U.S. production, within the capacity of the U.S. Army Air Forces to man and employ such aircraft.

Spare parts for the maintenance of these aircraft are not included in the total listed above, and adequate provision must be <u>added to these requirements.</u>

10. Total Personnel Requirements to meet this program in 1943.

Summary of Personnel Requirements to meet Programs A and B by January 1, 1944: (Including present on hand, and estimated attrition)

Table A3.6.

	Program A			Program B		
	Air Force	Services	Total	Air Force	Services	Total
Officers	230,243	72,600	302,843	253,000	86,260	339,260
E.M.	1,554,104	877,400	2,431,504	1,963,000	1,048,740	3,011,740
Total	1,784,347	950,000	2,734,347	2,216,000	1,135,000	3,351,000

Given the necessary priorities, these requirements can be met and trained. Of the 150,000 annual rate of aviation cadets estimated available, the A.A.F. will require 120,000 leaving 30,000 for the Navy.

11. Logistic Requirements.

	Table A3.7.	
	Program A	**Program B**
Bombs	1,140,363 tons	1,238,566 tons
Gasoline	4,888,941,000 gallons	5,372,179,000 gallons
Shipping required	17,421,507 ship tons	19,804,041 ship tons

Maximum number of 11,000 ton vessels required to be in use in any one month (average turn-around 2.81 months)

<u>Program A</u>   <u>Program B</u>
429   477

The Air Force requirements for shipping imposed by either program can be met, if the Navy requirements increase on a straight line basis applied against 1942 requirements, and if no other increase is made in Army strengths overseas beyond that attained by January 1, 1943.

The total gasoline requirement, is close to the maximum output that can be attained in the United Sates, using all productive facilities and without regard to any United States Navy or British requirements, <u>if all the gasoline is 100 octane.</u> The fact that a great deal of this gasoline can be 91 octane, for trainers, alleviates this situation to some extent.

12. Air Transport Command Requirements.

2,217 Transports are required, of which two-thirds should be long range, four engine.

13. Rates of Attrition.

It is likely that initial operations in the air offensives will be attended by an abnormally high rate of attrition. This may be expected as a result of losses in shipping caused by submarine operations before the air attack on submarine bases has taken effect, losses at bases before the attack of enemy bomber factories has taken effect, and losses from combat before the attack of fighter factories and attrition from air combat has reduced the enemy fighter forces. However, these loss rates should drop rapidly as our operations progress. It is believed that the rate of attrition of 20% per month from all causes in active combat zones will be fairly average. This is based upon British long-term experience.

14. Rates of Commitment.

In order to reach the increased goal of combat units by January 1944, it will probably be necessary to reduce the expected rate of commitment of units to combat theaters in early 1943, to increase the training establishments.

15. Conclusion.

   a. Both Germany and Japan are vulnerable to air attack.
   b. A successful air offensive against Germany can be carried out and is a necessary preliminary to ultimate victory over Germany.
   c. Base areas are now available in the United Kingdom, capable of sustaining the necessary air forces to accomplish this purpose.
   d. It is possible to conduct precision daylight bombing in the face of known and projected defenses of Western Europe.
   e. It is possible to conduct such an air offensive against Germany without prohibitive losses.
   f. Air support is essential to the conduct of all our other campaigns in 1943.
   g. It is possible to meet the logistic and personnel requirements for the air force necessary to gain victory over our enemy.
   h. It is believed possible to provide and deploy the necessary air forces in 1943 provided this requirement is given priority over all others including the allocation of necessary shipping, for an air offensive against Germany and support of land and sea forces in all other theaters.
   i. It is not believed possible to provide and deploy the necessary air forces in 1943 for simultaneous air offensives against Germany and Japan and air support of other essential operations.

Appendix 4
Combined Bomber Offensive Directive

Memorandum by the Combined Chiefs of Staff[1]
[Casablanca,] January 21, 1943

C.C.S. 166/1/D

THE BOMBER OFFENSIVE FROM THE UNITED KINGDOM

Directive to the appropriate British and U.S. Air Force Commanders, to govern the operation of the British and U.S. Bomber Commands in the United Kingdom (Approved by the Combined Chiefs of Staff at their 65th Meeting on January 21, 1943)

1. Your primary object will be the progressive destruction and dislocation of the German military, industrial and economic system, and the undermining of the morale of the German people to a point where their capacity for armed resistance is fatally weakened.

2. Within that general concept, your primary objectives, subject to the exigencies of weather and of tactical feasibility, will for the present be in the following order of priority:

    a. German submarine construction yards.
    b. The German aircraft industry.
    c. Transportation.
    d. Oil plants.
    e. Other targets in enemy war industry.

The above order of priority may be varied from time to time according to developments in the strategical situation. Moreover, other objectives of great importance either from the political or military point of view must be attacked. Examples of these are:

    (1) Submarine operating bases on the Biscay coast. If these can be put out of action, a great step forward will have been taken in the U-boat

war which the C.C.S. have agreed to be a first charge on our resources. Day and night attacks on these bases have been inaugurated and should be continued so that an assessment of their effects can be made as soon as possible. If it is found that successful results can be achieved, these attacks should continue whenever conditions are favorable for as long and as often as is necessary. These objectives have not been included in the order of priority, which covers long term operations, particularly as the bases are not situated in Germany.

(2) Berlin, which should be attacked when conditions are suitable for the attainment of specially valuable results unfavorable to the morale of the enemy or favorable to that of Russia.

3. You may also be required, at the appropriate time, to attack objectives in Northern Italy in connection with amphibious operations in the Mediterranean theater.

4. There may be certain other objectives of great but fleeting importance for the attack of which all necessary plans and preparations should be made. Of these, an example would be the important units of the German Fleet in harbor or at sea.

5. You should take every opportunity to attack Germany by day, to destroy objectives that are unsuitable for night attack, to sustain continuous pressure on German morale, to impose heavy losses on the German day fighter force, to contain German fighter strength away from the Russian and Mediterranean theaters of war.

6. When the Allied armies reenter the Continent, you will afford them all possible support in the manner most effective.

7. In attacking objectives in occupied territories, you will conform to such instructions as may be issued from time to time for political reasons by His Majesty's Government through the British Chiefs of Staff.

Appendix 5
Pointblank Directive

POINTBLANK DIRECTIVE

S.46368/A.C.A.S.(Ops.) 10th June, 1943

Sir,

I am directed to refer to Directive C.C.S.166/1/D dated 21st January, 1943, issued by the Combined Chiefs of Staff and forwarded to the Commanding General, Eighth Air Force and the Air Officer Commanding-in-Chief, Bomber Command under cover of Air Ministry letter S.46368/A.C.A.S.(Ops.) dated 4th February, 1943. This directive contained instructions for the conduct of the British and American bomber offensive from this country.

2. In paragraph 2 of the directive, the primary objectives were set out in order of priority, subject to the exigencies of weather and tactical feasibility. Since the issue of this directive there have been rapid developments in the strategical situation which have demanded a revision of the priorities originally laid down.

3. The increasing scale of destruction which is being inflicted by our night bomber forces and the development of the day bombing offensive by the Eighth Air Force have forced the enemy to deploy day and night fighters in increasing numbers on the Western Front. Unless this increase in fighter strength is checked we may find our bomber forces unable to fulfill the tasks allotted to them by the Combined Chiefs of Staff.

4. In these circumstances it has become essential to check the growth and to reduce the strength of the day and night fighter forces which the enemy can concentrate against us in this theatre. To this end the Combined Chiefs of Staff have decided that the first priority in the operation of British and

American bombers based in the United Kingdom shall be accorded to the attack of German fighter forces and the industry upon which they depend.

5. The primary object of the bomber forces remains as set out in the original directive issued by the Combined Chiefs of Staff (C.C.S.166/1/D dated 21st January, 1943) i.e.: "the progressive destruction and dislocation of the German military, industrial and economic system, and the undermining of the morale of the German people to a point where their capacity for armed resistance is fatally weakened."

6. In view, however, of the factors referred to in para. 4 of the following priority objectives have been assigned to the Eighth Air Force:

>Intermediate objective:
>>German Fighter Strength
>
>Primary objectives:
>>German submarine yards and bases
>>The remainder of the German aircraft industry
>>Ball bearings
>>Oil (contingent upon attacks against Ploesti from the Mediterranean)
>
>Secondary objectives:
>>Synthetic rubber and tyres
>>Military motor transport vehicles.

While the forces of the British Bomber Command will be employed in accordance with their main aim in the general disorganization of German industry their action will be designed as far as practicable to be complementary to the operations of the Eighth Air Force.

7. In pursuance of the particular requirements of para. 6 above, I am to request you to direct your forces to the following tasks:

>(i) the destruction of German airframe, engine and component factories and the ball-bearing industry of which the strength of the German fighter force depend.
>
>(ii) the general disorganization of those industrial areas associated with the above industries.
>
>(iii) the destruction of those aircraft repair depots and storage parks within range, and on which the enemy fighter force is largely dependent.

(iv) the destruction of enemy fighters in the air and on the ground.

The list of targets appropriate to these special tasks is in Appendix 'A' forwarded under cover of Air Ministry letter S.46368/3/D.B.Ops. dated 4th June, 1943. Further copies of this list, which will be amended from time to time as necessary, will be forwarded in due course.

8. Consistent with the needs of the air defence of the United Kingdom the forces of the British Fighter Command will be employed to further this general offensive by:
    (i) the attack of enemy aircraft in the air and on the ground.
    (ii) the provision of support necessary to pass bomber forces through the enemy defensive system with the minimum cost.

9. American fighter forces will be employed in accordance with the instructions of the Commanding General, Eighth Air Force in furtherance of the bomber offensive and in co-operation with the forces of Fighter Command.

10. The allocation of targets and the effective co-ordination of the forces involved is to be ensured by frequent consultation between the Commanders concerned. To assist this co-ordination a combined operational planning committee has been set up. The suggested terms of reference under which this Committee is to operate is outlined in Air Ministry letter CS.1936/A.C.A.S.(Ops) dated 10th June, 1943.

11. It is emphasized that the reduction of the German fighter force is of primary importance; any delay in its prosecution will make the task progressively more difficult. At the same time it is necessary to direct the maximum effort against the submarine construction yards and operating bases when tactical and weather conditions preclude attacks upon objectives associated with the German Fighter Force. The list of these targets is in Appendix 'B' forwarded with the Appendix 'A' referred to in paragraph 7 above.

I am, Sir,
Your obedient Servant,
(Sgd) N. H. Bottomley
Air Vice Marshal, A.C.A.S. (Ops).

Notes

Introduction

1. Lowell Getz, "Return From Bremen: The Low Squadron Is Gone," in *"Mary Ruth" Memories of Mobile . . . We Still Remember: Stories from the 91st Bombardment Group (Heavy), 8th Air Force, World War II*, 2001, http://www.91stbombgroup.com/mary_ruth/Chapter_3.htm (accessed May 28, 2018); Kit Carter and Robert Mueller, *U.S. Army Air Forces in World War II: Combat Chronology, 1941–1945* (Washington, DC: Center for Air Force History, 1991), 142; 107 of the 115 bombers dropped their bombs on the target area. Eighth Air Force, "Bomber Command Narrative of Operations April 17, 1943, Mission No. 52," Air Force Historical Research Agency, Maxwell AFB, Montgomery, AL (hereafter AFHRA), 520.332; 91st Bombardment Group Official Diary, April 1943, AFHRA, GP-91-HI 1, 31. Other groups included the 303rd Bombardment Group from RAF Molesworth, the 305th Bombardment Group from RAF Chelveston, and the 306th Bombardment Group from RAF Thurleigh. Robert Freeman, *Mighty Eighth War Diary* (Minneapolis: Motorbooks, 1970).

2. John Kreis, *Piercing the Fog* (Washington, DC: Air Force History and Museums Program, 1996), 140; Getz, "Return from Bremen"; Wilbur Morrison, *The Incredible 305th: The "Can Do" Bombers of World War II* (New York: Berkley, 1962), 52, 57. Four missions were flown in January, three in February, six in March, and four in April 1943: 306 Bombardment Group War Diaries, www.306bg.us/group_diary/jan43gpdiary.pdf, www.306bg.us/group_diary/feb43gpdiary.pdf, www.306bg.us/group_diary/mar43gpdiary.pdf, www.306bg.us/group_diary/apr43gpdiary.pdf, (accessed May 28, 2017).

3. John Craven, ed., *The 305th Bomb Group in Action: An Anthology* (Burleson, TX: 305th Bombardment Group (H) Memorial Association, 1990), 119; Walter Thom, *The Brotherhood of Courage: The History of the 305th Bombardment Group (H) in World War II* (New York: Martin Cook, 1986), 80–81.

4. Eighth Air Force, "Bomber Command Narrative of Operations 17 April, 1943, Mission No. 52," AFHRA, 520.332; United States Strategic Bombing Survey (USSBS), *Aircraft Division Airframes Plant Report No. 5: Focke-Wulf Aircraft Inc, Bremen, Germany*, September 26, 1945, Exhibit D, Bomb Plot, Bremen Airport, April 17, 1943; Getz, "Return from Bremen."

5. All 16 aircraft lost came from the lead formation. 91st Bombardment Group Official Diary, 31–32; Morrison, *The Incredible 305th*, 52, 57, 59; Getz, "Return from Bremen."

6. Kreis, *Piercing the Fog*, 140; Wesley Craven and James Cate, eds., *The Army Air Forces in World War II*, vol. 1, *Plans and Early Operations, January 1939 to August 1942* (Chicago: University of Chicago Press, 1948), 330. The B-17G added an extra chin gun to reduce this vulnerability. The German fighters also descended upon the low bomber group (306th) because it had difficulty maintaining a tight, mutually supportive formation The 306th Bomb group lost 10 out of the 16 bombers in its formation, 306th Bombardment Group History, October 20, 1943, March 1942–June 1945, AFHRA, GP-306-HI.

7. In all, 80 were killed, 92 interred, and 19 evaded. Kreis, *Piercing the Fog*, 140. Wyler went on to produce the documentary *The Memphis Belle: A Story of a Flying Fortress*. The *Memphis Belle*, also from the 91st Bombardment Group, completed its 19th mission against Bremen on April 17, 1943. Getz, "Return from Bremen"; 303rd Bomb Group (H), "Hells Angels vs. Memphis Belle," www.303rdbg.com/h-ha-mb.html (accessed May 28, 2018). The next mission over Germany was May 14, 1943, against submarine yards at Kiel. Eighth Air Force Combat Diary, www.8thafhs.org/combat1943.htm (accessed May 28, 2018); Clayton Chun, *Aerospace Power in the Twenty-First Century: A Basic Primer* (Colorado Springs, CO: United States Air Force Academy, 2001), 122.

8. John Weal, *Defence of the Reich Aces* (New York: Bloomsbury, 2012). The largest of the factories had been moved to Marienburg, in East Prussia near the Czechoslovakian border. Kreis, *Piercing the Fog*, 140; USSBS, *Focke-Wulf Aircraft Inc.*, 3, 12.

9. This was not the first use of weapons by aircraft. The Italians in their campaign against the Ottoman Empire in Libya in 1911 added weapons to their aircraft.

10. H. A. Jones, *The War in the Air: Being the Story of the Part Played in the Great War by the Royal Air Force*, vol. 3 (Oxford: Clarendon, 1931), 90, 105, 145, 243.

11. Andrew Boyle, *Trenchard* (London: Collins, 1962), 221, 226–27, 271; Smuts Report, www.rafmuseum.org.uk/london/whats-going-on/news/read-the-smuts-report/ (accessed May 28, 2017).

12. George Williams, "The Shank of the Drill: Americans and Strategical Aviation in the Great War," *Journal of Strategic Studies* 19, no. 3 (September 1996): 386–67; the Gorrell strategic bombing plan is available in Maurer Maurer, *The U.S. Air Service in World War I*, vol. 2 (Washington, DC: Office of Air Force History, 1978), 141–51.

13. Boyle, *Trenchard*, 311–12; Tami Davis Biddle, *Rhetoric and Reality in Air Warfare: The Evolution of British and American Ideas about Strategic Bombing, 1914–1945* (Princeton: Princeton University Press, 2002), 42.

14. Biddle, *Rhetoric and Reality*, 57.

15. Biddle, *Rhetoric and Reality*, 61.

16. Maurer, *The U.S. Air Service in World War I*, 501.

17. Richard Simpkin, *Deep Battle: The Brainchild of Marshal Tukhachevsky* (London: Brassey's Defence, 1987); Heinz Guderian, *Panzer Leader* (New York: Da Capo, 1952); Erwin Rommel, *The Rommel Papers* (New York: Da Capo, 1953). In Great Britain an exception is J. C. Slessor, *Air Power and Armies* (Oxford: Oxford University Press, 1936).

18. Giulio Douhet, *Command of the Air* (Washington, DC: Air Force History and Museums Program, 1998).

19. Douhet, *Command of the Air*, 16–17, 29, 31, 45, 56, 57, 116–17.

20. Douhet, *Command of the Air*, 5, 10, 25, 58, 60, 98, 128, 196.

21. Phillip Meilinger, *The Paths of Heaven: The Evolution of Airpower Theory* (Maxwell AFB, AL: Air University Press, 1997), 20–30.
22. It would take German intervention to return Mussolini to power.
23. Meilinger, *Paths of Heaven,* 33; Alfred Hurley, *Billy Mitchell: Crusader for Air Power* (Bloomington: Indiana University Press, 1975), 75.
24. Hugh Trenchard, "Memorandum by the Chief of the Air Staff for the Chiefs of Staff Sub Committee on the Object of an Air Force, 2nd May 1928," in *The Strategic Air Offensive against Germany, 1939–1945,* vol. 4, *Annexes and Appendices,* ed. Charles Webster and Noble Frankland (London: Her Majesty's Stationery Office, 1961), appendix 2, 71.
25. Hugh Trenchard, "Report on the Independent Air Force," *Tenth Supplement to London Gazette* January 1, 1919, 134–35.
26. Stanley Baldwin, "Mr Baldwin on Aerial Warfare—A Fear for the Future," *Times,* November 11, 1932, 7.
27. Daniel Hucker, *Public Opinion and the End of Appeasement in Britain and France* (Surrey, UK: Ashgate, 2011), 69.
28. Butt Report, August 18, 1941, found in British National Archives, AIR 14/1218, or view online at https://etherwave.files.wordpress.com/2014/01/butt-report-transcription-tna-pro-air-14-12182.pdf (accessed May 28, 2018).
29. For a biography of Billy Mitchell, see Hurley, *Billy Mitchell.* Mark Clodfelter, *Beneficial Bombing: The Progressive Foundations of American Air Power, 1917–1945* (Lincoln: University of Nebraska Press, 2010), 15.
30. Hurley, *Billy Mitchell,* 43.
31. Hurley, *Billy Mitchell,* 73.
32. William Mitchell, "Notes on the Multi-motored Bombardment Group Day and Night," 1923, Fairchild Special Collection, 358.4 U58n, Air University Library, Maxwell AFB, Montgomery, AL.
33. Mitchell, "Notes on the Multi-motored Bombardment Group," 84.
34. Mitchell, "Notes on the Multi-motored Bombardment Group," 76.
35. Mitchell, "Notes on the Multi-motored Bombardment Group," 93.
36. Mitchell, "Notes on the Multi-motored Bombardment Group," 94.
37. Hurley, *Billy Mitchell,* 86–88.
38. Hurley, *Billy Mitchell,* 95, 98.
39. William Mitchell, *Winged Defense: The Development and Possibilities of Modern Air Power, Economic and Military* (New York: Dover, 1988); Hurley, *Billy Mitchell,* 101, 107.
40. Robert Finney, *History of the Air Corps Tactical School, 1920–1940* (Maxwell AFB, AL: USAF Historical Division, 1955), 10, 14.
41. Thomas Greer, *The Development of Air Doctrine in the Army Air Arm, 1917–1941,* USAF Historical Study 89 (Maxwell AFB, AL: Air University, 1955), 29.
42. Sherman had edited "Tactical History of the Air Service in World War I" under the direction of Edgar Gorrell. He then wrote "Tentative Manual for the Employment of the Air Service" in 1919. William Sherman, *Air Warfare* (Maxwell AFB, AL: Air University Press, 2002), iii, x.
43. Sherman, *Air Warfare,* 19, 24.
44. Sherman, *Air Warfare,* 194, 196–97.
45. Finney, *History of the Air Corps Tactical School,* 18–19, 23–25.

46. Finney, *History of the Air Corps Tactical School*, 21.

47. Greer, *The Development of Air Doctrine in the Army Air Arm*, 39, 46; Gordon Williams, "A Progressive History of the Air Corps Egg Layers" *Flying Magazine* 19, no. 5 (November 1936): 15.

48. Albert Pardini, *The Legendary Norden Bombsight* (Atglen, PA: Schiffer, 1999), 45, 67, 72, 74; Stephen McFarland, *America's Pursuit of Precision Bombing, 1910–1945* (Washington, DC: Smithsonian Institute Press, 1995), 69.

49. Pardini, *The Legendary Norden Bombsight*, 77, 87, 96.

50. Despite a highly anticipated rollout, however, the B-17's start was rather inauspicious when, in trials, the single test aircraft crashed on takeoff due to pilot error. Peter Bowers, *Boeing Aircraft since 1916* (New York: Funk & Wagnalls, 1966), 245–46; Kenneth Munson, *Aircraft of World War II* (New York: Doubleday, 1962), 32.

51. Later B-17 models with turbocharged engines reached speeds of 295 mph at 25,000 feet and a service ceiling of 35,000 feet. "B-17 Aircraft," *Encyclopedia Britannica*, www.britannica.com/technology/B-17 (accessed May 28, 2018); Bowers, *Boeing Aircraft since 1916*, 256; McFarland, *America's Pursuit of Precision Bombing*, 94–95.

52. McFarland, *America's Pursuit of Precision Bombing*, 94–95.

53. Finney, *History of the Air Corps Tactical School*, 39–40.

54. Finney, *History of the Air Corps Tactical School*, 40.

55. Finney, *History of the Air Corps Tactical School*, 42–43, 54–55.

56. Finney, *History of the Air Corps Tactical School*, 37.

1. Air Power and War

1. Lieutenant General Harold L. George official USAF biography, www.af.mil/About-Us/Biographies/Display/Article/107023/lieutenant-general-harold-1-george/ (accessed May 28, 2018).

2. Harold George credited this insight to his reading of B. H. Liddell Hart's *The Real War 1914–1918* (Boston: Little, Brown, 1930). See Murray Green, "Interview, LtGen Harold L. George, West Los Angeles, California, March 16, 1970," micfilm 43821, IS 168.7326-169, 25, Air Force Historical Research Agency, Maxwell AFB, AL (hereafter AFHRA).

3. Harold George, "An Inquiry into the Subject 'War,'" Air Force Course, Air Corps Tactical School, 1936, AFHRA, 248.11-9, 2.

4. Henry Wadsworth Longfellow, "A Psalm of Life: What the Heart of the Young Man Said to the Psalmist" (1838).

5. J. F. C. Fuller, *The Reformation of War* (London: Hutchinson, 1923).

6. Carl Von Clausewitz, *On War* (Princeton: Princeton University Press, 1976), 87.

7. Fuller, *The Reformation of War*, 1.

2. The Objective of Air Warfare

1. General Muir Fairchild official USAF biography, www.af.mil/About-Us/Biographies/Display/Article/107112/general-muir-s-fairchild/ (accessed May 28, 2018).

2. Muir Fairchild, "Air Power and Air Warfare," Air Force Course, Air Corps Tactical School, Air Force Historical Research Agency, Maxwell AFB, AL (hereafter AFHRA), 1939, 248.2020A-1.

Notes to Pages 54–87   275

3. H. A. Jones, *The War in the Air: Being the Story of the Part Played in the Great War by the Royal Air Force*, vol. 3 (Oxford: Clarendon, 1931), 243–48.

4. R. K. Turner, "Employment of Aviation in Naval Warfare" (Newport, RI: US Naval War College Department of Operations, December 1, 1936), 12, https://usnwcarchive.org/files/original/f913449f101790247b96fbb6f00b2e08.pdf (accessed May 28, 2018).

5. Turner, "Employment of Naval Aviation in Naval Warfare," 12–13.

6. Giulio Douhet, *Command of the Air* (Washington, DC: Air Force History and Museums Program, 1998), 29, 31, 56, 57.

7. Murray Williamson, *German Military Effectiveness* (Mt. Pleasant, SC: Nautical & Aviation, 1992), 64.

8. Brigadier General Donald Wilson official USAF biography, www.af.mil/About-Us/Biographies/Display/Article/108688/brigadier-general-donald-wilson/ (accessed May 28, 2018).

9. Donald Wilson, "Principles of War" Air Force Course, Air Corps Tactical School, 1939, AFHRA, 248.2021A-2.

10. *Field Service Regulations*, part 1, *Operations, 1909* (London: HMSO, 1914), 14.

11. Marshal Marmont was a French general who betrayed his close friend Napoleon and later wrote a thesis on war. Auguste Frédéric Louis Viesse de Marmont, Duc de Ragusa, *The Spirit of Military Institutions; or, Essential Principles of the Art of War*, trans. Henry Coppee (Philadelphia: Lippincott, 1862), 36.

12. The Battle of Crecy, August 26, 1346, ended with the victory of England over France.

13. The Battle of Agincourt, October 25, 1415, was an English victory over France.

14. Major General George A. Lynch was chief of infantry from May 24, 1937, to April 30, 1941, www.westpointaog.org/memorial-article?id=be3913d9-bc2e-4cd8-a44d-702e6ad30cef (accessed May 28, 2018).

15. Thomas Phillips, "Word Magic of the Military Mystics," *Coastal Artillery Journal*, September–October 1939, 429–30.

16. Harold George, "Principles of War," Air Force Course, Air Corps Tactical School, 1935–1936, AFHRA, 248.2017A-9.

17. Major General Haywood S. Hansell official USAF biography, www.af.mil/About-Us/Biographies/Display/Article/106813/major-general-haywood-s-hansell-jr/ (accessed May 28, 2018).

18. Haywood Hansell, "The Aim in War," Air Force Course, Air Corps Tactical School, 1936–1937, AFHRA, 248.2018A-3.

19. Basil Liddell-Hart, *The Decisive Wars of History* (London: G. Bell & Sons, 1929), 148.

20. Frank Simonds and Brooks Emeny, *International Relations and Economic Nationalism* (New York: American Book, 1935), 154.

21. Alexander Pope, *An Essay on Criticism* (London: W. Lewis, 1711).

3. The Bomber Always Gets Through

1. Brigadier General Kenneth Walker official USAF biography, www.af.mil/About-Us/Biographies/Display/Article/105285/brigadier-general-kenneth-newton-walker/ (accessed May 28, 2018).

2. Air Corps Tactical School, *Bombardment Aviation* (February 1931), Air Force Historical Research Agency, Maxwell AFB, AL (hereafter AFHRA), 248.101-9.

3. Kenneth Walker, "Driving Home the Bombardment Attack," *Coast Artillery Journal* 73, no. 4 (October 1930): 328–40.

4. Major General Frederick M. Hopkins official USAF biography, http://www.af.mil/About-Us/Biographies/Display/Article/106713/major-general-frederick-m-hopkins-jr/.

5. Frederick Hopkins, "Tactical Offense and Tactical Defense," Air Force Course, Air Corps Tactical School, 1939, AFHRA, 248.2020A-5. Major Charles E. Thomas served on the ACTS faculty. See www.af.mil/About-Us/Biographies/Display/Article/105444/major-general-charles-e-thomas-jr/ (accessed May 28, 2018).

6. Lieutenant Colonel Baron Von Der Goltz, *The Nation in Arms*, trans. Philip A. Ashworth (London: W. H. Allen, 1887).

7. Lieutenant Colonel Baron Von Der Goltz, *The Nation in Arms*, 5.

8. Claire Chennault resigned from the Air Corps in April 1937 and shortly thereafter traveled to China to consult for the Chinese Air Force. Martha Byrd, *Chennault: Giving Wings to the Tiger* (Tuscaloosa: University of Alabama Press, 1987).

9. Captain Townsend Griffiss was an Air Corps pilots assigned to the Spanish Civil War as an observer. He became a student at the ACTS in 1938. In 1942 he was killed when an RAF B-24 he was a passenger in was shot down over the English Channel, making him the first American aviator killed in the European theater in World War II. Griffiss Air Force Base is named after him. Stephen Mulvey, "Townsend Griffiss, Forgotten Hero of World War II," *BBC News Magazine*, February 14, 2012, http://www.bbc.com/news/magazine-17011105 (accessed May 28, 2018).

4. High-Altitude Daylight Precision Bombardment

1. General Laurence Kuter official USAF biography, www.af.mil/About-Us/Biographies/Display/Article/106523/general-laurence-s-kuter/ (accessed May 29, 2018).

2. Bombing accuracy is measured by the mean probable error, which is defined as the distance from the target within which half of bombs that are dropped are expected to fall.

3. Laurence Kuter, "Practical Bombing Probabilities," Bombardment Aviation, Air Corps Tactical School, Air Force Historical Research Agency, Maxwell AFB, AL (hereafter AFHRA), 1939, 248.2208B-7pt4.

4. Air Corps Tactical School, *Bombardment* November 1935, 54, AFHRA, 248.101-9.

5. http://www.af.mil/About-Us/Biographies/Display/Article/106859/lieutenant-general-hubert-reilly-harmon/ (accessed May 28, 2018).

6. Colonel Edgar P. Sorensen served as director of the Air Corps Board and then commandant of ACTS. http://www.af.mil/About-Us/Biographies/Display/Article/108681/brigadier-general-edgar-p-sorensen/ (accessed May 28, 2018).

7. The author was required to calculate mean probable errors during his A-10 weapons instructor course in 1997. Only the large-scale employment of precision-guided bombs reduced the importance of such calculations.

5. Vital and Vulnerable

1. Muir Fairchild, "National Economic Structure," Air Force Course, Air Corps Tactical School, Air Force Historical Research Agency, Maxwell AFB, AL (hereafter AFHRA), 1939, 248.2020A-9.

2. J. F. C. Fuller, *The Reformation of War* (London: Hutchinson, 1923) 87.

3. On June 15, 2015, a Zeppelin raid struck the Palmer Works at Jarrow, England. "German Zeppelins: Terrorizing the British and RAF during WW1," *Military History*, September 3, 2015, http://warfarehistorynetwork.com/daily/military-history/german-zeppelins-terrorizing-the-british-and-raf-during-ww1/ (accessed May 29, 2018).

4. Federal Power Policy Commission, *Interim Report National Power Survey*, March 15, 1935.

5. Major General Edward M. Markham served as chief of the Army Corps of Engineers from 1933 to 1937. See US Army Corps of Engineers History, www.hq.usace.army.mil/history/coe3.htm (accessed May 29, 2018).

6. Muir Fairchild, "New York Industrial Area," Air Force Course, Air Corps Tactical School, 1939, AFHRA, 248.2020A-11.

7. United States Strategic Bombing Survey, *Summary Report* (*European War*) (Maxwell AFB, AL: Air University Press, 1987), 39.

6. What to Target

1. Muir Fairchild, "Primary Strategic Objectives of Air Forces," Air Corps Tactical School Lectures 1938–1939, Air Force Historical Research Agency, Maxwell AFB, AL, 248.2020A-14.

7. High-Altitude Daylight Precision Bombing in World War II

1. Robert T. Finney, *History of the Air Corps Tactical School, 1920–1940* (Maxwell AFB, AL: USAF Historical Division, 1955), 79; H. H. Arnold, "Circular Letter No. 39-11: Instituted Abbreviated Courses at the Air Corps Tactical School," April 3, 1939, Air Force Historical Research Agency, Maxwell AFB, AL (hereafter AFHRA), 248.2021-1A PT.2, IRIS 00159387.

2. Bomber Command suffered over 55,000 deaths and Eighth Air Force over 26,000. Bomber Command Museum of Canada, www.bombercommandmuseum.ca/commandlosses.html (accessed November 7, 2017); U.S. Air Force Fact Sheet, "Eighth Air Force History," www.8af.af.mil/About-Us/Fact-Sheets/Display/Article/333781/eighth-air-force/ (accessed May 30, 2018).

3. Some 40 percent of bombers earmarked for Eighth Air Force were diverted to the Mediterranean. Haywood Hansell, *Strategic Air War against Germany and Japan* (Washington, DC: Office of Air Force History, 1986), 261.

4. The Phoney War was the eight-month period from France and Britain's declaration of war in September 1939 after Germany's invasion of Poland until Germany's invasion of France in May 1940. It is called the Phoney War because of France and Britain's minimal military activity.

5. Tami Davis Biddle, *Rhetoric and Reality in Air Warfare: The Evolution of British and American Ideas about Strategic Bombing, 1914–1945* (Princeton: Princeton University Press, 2002), 183–84.

6. Stephen Bungay, *The Most Dangerous Enemy: A History of the Battle of Britain* (London: Aurum, 2001) 368–88.

7. Biddle, *Rhetoric and Reality*, 191.

8. Biddle, *Rhetoric and Reality*, 194; Butt Report, August 18, 1941, British National Archives, AIR 14-1218, https://etherwave.files.wordpress.com/2014/01/butt-report-transcription-tna-pro-air-14-12182.pdf (accessed May 30, 2018).

9. James Gaston, *Planning the American Air War* (Washington, DC: National Defense University Press, 1982), 11.

10. Haywood S. Hansell, *The Air Plan That Defeated Hitler* (Atlanta: Higgins-McArthur, 1972), 31.

11. Gaston, *Planning the American Air War*, 90.

12. Hansell, *The Air Plan That Defeated Hitler*, 32; Mark Clodfelter, *Beneficial Bombing: The Progressive Foundations of American Air Power, 1917–1945* (Lincoln: University of Nebraska Press, 2010), 93.

13. Hansell, *The Air Plan That Defeated Hitler*, 31; Clodfelter, *Beneficial Bombing*, 91. Aluminum was particularly important for aircraft construction.

14. Hansell, *The Air Plan That Defeated Hitler*, 36.

15. Hansell, *The Air Plan That Defeated Hitler*, 38.

16. Hansell, *The Air Plan That Defeated Hitler*, 58.

17. Biddle, *Rhetoric and Reality*, 197; Phillips O'Brien, *How the War Was Won* (Cambridge: Cambridge University Press, 2015), 273.

18. Biddle, *Rhetoric and Reality*, 211; Hansell, *The Air Plan That Defeated Hitler*, 93; O'Brien, *How the War Was Won*, 274.

19. On December 31, 1942, the USAAF formed the Committee of Operations Analysis to make recommendations on strategic targeting. Charles Shrader, *History of Operations Research Analysis in the United States Army*, vol. 1, *1942–1962* (Washington, DC: United States Army, 2006), 25.

20. Adam Tooze, *Wages of Destruction: The Making and Breaking of the Nazi Economy* (New York: Penguin, 2006), 600–601; O'Brien, *How the War Was Won*, 283.

21. Ball bearings were deemed a vital and vulnerable component of aircraft engine production. O'Brien, *How the War Was Won*, 277.

22. Hansell, *The Air Plan That Defeated Hitler*, 86; O'Brien, *How the War Was Won*, 281.

23. Biddle, *Rhetoric and Reality*, 232; O'Brien, *How the War Was Won*, 284.

24. Bomber Command lost over 55,000 aircrew from September 1939 to May 1945. See www.bombercommandmuseum.ca/commandlosses.html (accessed May 30, 2018). Eighth Air Force, operating from England, lost over 26,000: U.S. Air Force Fact Sheet: Eighth Air Force History, www.8af.af.mil/About-Us/Fact-Sheets/Display/Article/333794/eighth-air-force-history/ accessed May 30, 2018).

25. Clodfelter, *Beneficial Bombing*, 130; W. Hays Parks, "'Precision' and 'Area' Bombing: Who Did Which, and When?" *Journal of Strategic Studies* 18, no. 1 (1995): 148, 150.

26. Parks, "'Precision' and 'Area' Bombing," 151, 154. Interestingly, it was the opposite for Bomber Command as cloud cover over Germany tends to dissipate at night during the winter months. Clodfelter, *Beneficial Bombing*, 151.

27. Clodfelter, *Beneficial Bombing*, 137.

28. O'Brien, *How the War Was Won*, 282, 288.

29. Clodfelter, *Beneficial Bombing*, 119–20.

30. Biddle, *Rhetoric and Reality*, 226. Most of the B-24 bombers, considered more vulnerable than the B-17, were deployed to the Mediterranean theater.

31. Italy's Regia Aeronautica also embraced strategic bombing, but did not have the industrial capacity to produce the quality and quantity of bombers to operate a heavy bomber force.

32. Clodfelter, *Beneficial Bombing*, 155.

33. Williamson Murray, *Strategy for Defeat: The Luftwaffe, 1933–45* (Honolulu: University Press of the Pacific, 2002), 227.

34. Murray, *Strategy for Defeat*, 273–74.

35. Woody Parramore, "The Combined Bomber Offensive's Destruction of Germany's Refined-Fuels Industry" *Air & Space Power Journal*, March–April 2012, 77, 79–82; Clodfelter, *Beneficial Bombing*, 165–66, 171–72; Robert Pape, *Bombing to Win* (Ithaca, NY: Cornell University Press, 1996), 282. Pape argues that the Soviet Red Army's overrunning of Ploesti oil fields in Romania (August 1944) and Hungary (February 1945) caused the collapse of German oil production. However, by then the German Army and Air Force could not utilize imported fuel supplies as USAAF had repeatedly attacked Ploesti prior to the Soviets' arrival, and by the summer of 1944 the German Army had become reliant on domestic sources of fuel; the Luftwaffe required the domestically produced synthetic high-octane fuel for its fighters.

36. Railways, rolling stock, and bridges were primarily struck by USAAF medium bombers and fighter bombers, while heavy bombers conducted HADPB and radar attacks against French railway marshalling yards. Horst Boog, Gerhard Krebs, and Detlef Vogel, eds., *Germany and the Second World War*, vol. 7, *The Strategic Air War in Europe and the War in the West and East Asia, 1943–1944/5* (Oxford: Clarendon, 2006), 143, 135; Clodfelter, *Beneficial Bombing*, 161; Murray, *Strategy for Defeat*, 267–71.

37. Boog, Krebs, and Vogel, *Germany and the Second World War*, 509–10; Murray, *Strategy for Defeat*, 263; Dieter Ose, "Rommel and Rundstedt: The 1944 Panzer Controversy," *Military Affairs* 5, no. 1 (January 1986): 7–11.

38. Kenneth P. Werrell, "The Strategic Bombing of Germany in World War II: Costs and Accomplishments," *Journal of American History* 73, no. 3 (December 1986): 703–13.

39. Richard Overy, *Why the Allies Won* (New York: Norton, 1995), 102–3; Richard Overy, *The Bombers and the Bombed: Allied Air War over Europe, 1940–1945* (New York: Viking, 2013), 102–3; O'Brien, *How the War Was Won*, 141; Jeremy Black, *Introduction to Global Military History: 1775 to the Present Day* (London: Routledge, 2006), 154.

40. Stephen W. Twing, *Myths, Models & U.S. Foreign Policy: The Cultural Shaping of Three Cold Warriors* (London: Lynne Rienner, 1998), 104.

41. Martin Gilbert, "Churchill Proceedings—Churchill and Bombing Policy," Fifth Churchill Lecture, George Washington University, October 18, 2005, www.winstonchurchill.org/publications/finest-hour/finest-hour-137/churchill-proceedings-churchill-and-bombing-policy/ (accessed May 30, 2018).

42. Timothy Johnston, *Being Soviet: Identity, Rumour, and Everyday Life under Stalin, 1939–1953* (Oxford: Oxford University Press, 2011), 51.

43. Vojtech Mastny, "Stalin and the Prospects of a Separate Peace in World War II," *American Historical Review* 77, no. 5 (December 1972): 1369, 1366.

44. Mastny, "Stalin," 1375; F. C. Jones, *Japan's New Order in East Asia: Its Rise and Fall, 1937–45* (New York: AMS, 1978), 413.

45. Winston S. Churchill, *Road to Victory* (London: Heinemann, 1966), 430; Mastny, "Stalin," 1378.

46. Mastny, "Stalin," 1379.
47. Michael S. Sherry, *The Rise of American Air Power* (New Haven: Yale University Press, 1987), 158.
48. Tooze, *Wages of Destruction*, 596–600.
49. Boog, Krebs, and Vogel, *Germany and the Second World War*, 26.
50. R. V. Jones, *The Wizard War: British Scientific Intelligence, 1939–1945* (New York: Coward, McCann & Geoghegan, 1978).
51. Martin Bowman, *De Havilland Mosquito* (Ramsbury, UK: Crowood, 2005); Werrell, "The Strategic Bombing of Germany in World War II," 704.
52. Harry Holmes, *Avro Lancaster: The Definitive Record* (London: Butler & Tanner, 1997).
53. Boog, Krebs, and Vogel, *Germany and the Second World War*, 69; Murray, *Strategy for Defeat*, 170. An exception to Eighth Air Force bombing aircraft production was its decision in late July 1943 to join Bomber Command in Operation Gomorrah, the bombing of Hamburg.
54. Overy, *The Bomber and the Bombed*, 204–5.
55. Clodfelter, *Beneficial Bombing*, 182.
56. O'Brien, *How the War Was Won*, 290.
57. Richard Muller, *The German Air War in Russia* (Baltimore, MD: Nautical & Aviation, 1992), 158.
58. O'Brien, *How the War Was Won*, 305.
59. This includes the Eastern Front, Italy, Balkans, and Norway/Denmark. O'Brien, *How the War Was Won*, 306. Not only were these guns unavailable for air defense duty on the Eastern Front, but being dual-use weapons they could also be employed against ground forces. The German "eighty-eight," an 88-millimeter cannon, produced by the tens of thousands, proved especially lethal against Soviet armor.
60. Murray, *Strategy for Defeat*, 190.
61. O'Brien, *How the War Was Won*, 305.
62. O'Brien, *How the War Was Won*, 291; Muller, *The German Air War in Russia*, 145.
63. Michael Neufeld, "Hitler, the V-2, and the Battle for Priority, 1939–1943," *Journal of Military History* 57, no. 3 (July 1993): 515, 531–32; Roy Irons, *Hitler's Terror Weapons: The Price of Vengeance* (London: HarperCollins, 2002), 31; Tooze, *Wages of Destruction*, 619; Williamson Murray, "Reflections on the Combined Bomber Offensive," *Militargeschichtliche Mitteilungen* 51 (1992): 90; O'Brien, *How the War Was Won*, 305, 306–7; Murray, *Strategy for Defeat*, 189.
64. Kenneth Werrell, *Blankets of Fire: U.S. Bombers over Japan during World War II* (Washington, DC: Smithsonian Institute Press, 1996); Walter Boyne, "The B-29's Battle of Kansas," *Air Force Magazine*, February 2012, 95; O'Brien, *How the War Was Won*, 47.
65. Clodfelter, *Beneficial Bombing*, 195.
66. Boyne, "The B-29's Battle of Kansas," 94–95.
67. Hansell, *The Strategic Air War against Germany and Japan*, 190; Clodfelter, *Beneficial Bombing*, 206.
68. Hansell, *The Strategic Air War against Germany and Japan*, 159, 153 (other basing options considered were Alaska, the Philippines, Formosa, and Okinawa), 167; Clodfelter, *Beneficial Bombing*, 192.
69. Boyne, "The B-29's Battle of Kansas," 94, 97.

70. Boyne, "The B-29's Battle of Kansas," 97; Clodfelter, *Beneficial Bombing*, 200.

71. Hansell, *The Strategic Air War against Germany and Japan*, 167, 189; Clodfelter, *Beneficial Bombing*, 205–6.

72. Clodfelter, *Beneficial Bombing*, 207. Norstad effectively ran 20th Air Force while Arnold recovered from a fourth heart attack. Hansell, *The Strategic Air War against Germany and Japan*, 199.

73. Clodfelter, *Beneficial Bombing* 219; Sherry, *The Rise of American Air Power*, 277; Hansell, *The Strategic Air War against Germany and Japan*, 222; Biddle, *Rhetoric and Reality*, 269.

74. For an assessment of the Japanese surrender, see S.C.M. Paine, *The Wars of Asia, 1911–49* (New York: Cambridge University Press, 2012).

75. Clodfelter, *Beneficial Bombing*, 226.

76. Clodfelter, *Beneficial Bombing*, 232; Hansell, *The Strategic Air War against Germany and Japan*, 198–200.

77. Clodfelter, *Beneficial Bombing*, 232.

78. Paine, *The Wars of Asia*, 209–11.

79. The advent of nuclear weapons has not led air power to win wars, but perhaps it has contributed to an even more important outcome: the avoidance of wars through deterrence.

Appendix 4

1. Foreign Relations of the United States, the Conferences at Washington, 1941–1942, and Casablanca, 1943, document 412, https://history.state.gov/historicaldocuments/frus1941-43/d412 (accessed May 30, 2018).

Bibliography

Ahman, Hugh. "Interview of Maj Gen Donald Wilson." Carmel, CA, December 10–11, 1975. United States Air Force Oral History Program AFHRA K239.0512.878.
Anthony, Victor. "The United States Army Air Corps, 1926–1939." In *Modern Warfare and Society*. Maxwell AFB, AL: Air Command and Staff College, 1989.
Barker, J. D. *History of the Air Corps Tactical School*. Maxwell AFB, AL: USAF Historical Research Center, 1931.
Benton, Jeffrey C. *They Served Here: Thirty-Three Maxwell Men*. Maxwell AFB, AL: Air University Press, 1999.
Biddle, Tami Davis. *Rhetoric and Reality in Air Warfare: The Evolution of British and American Ideas about Strategic Bombing, 1914–1945*. Princeton: Princeton University Press, 2002.
Black, Jeremy. *Introduction to Global Military History: 1775 to the Present Day*. London: Routledge, 2006.
Boog, Horst, Gerhard Krebs, and Detlef Vogel, eds. *Germany and the Second World War*. Vol. 7: *The Strategic Air War in Europe and the War in the West and East Asia, 1943–1944/5*. Oxford: Clarendon, 2006.
Bowers, Peter M. *Boeing Aircraft since 1916*. New York: Funk & Wagnalls, 1966.
———. *Fortress of the Sky*. Grenade Hills, CA: Sentry Books, 1976.
Bowman, Martin. *De Havilland Mosquito*. Ramsbury, UK: Crowood, 2005.
———. *We Were Eagles: The Eighth Air Force at War, July 1942 to November 1943*. Vol. 1. Gloucestershire, UK: Amberley, 2014.
Boyle, Andrew. *Trenchard*. London: Collins, 1962.
Boyne, Walter, ed. *Air Warfare: An International Encyclopedia*. Santa Barbara, CA: ABC-CLIO, 2002.
———. "The B-29's Battle of Kansas." *Air Force Magazine*, February 2012.
"B-17 Aircraft." *Encyclopedia Britannica*. Accessed May 30, 2018, www.britannica.com/technology/B-17.
Bungay, Stephen. *The Most Dangerous Enemy: A History of the Battle of Britain*. London: Aurum, 2001.
Byrd, Martha. *Chennault: Giving Wings to the Tiger*. Tuscaloosa: University of Alabama Press, 1987.
———. *Kenneth N. Walker: Airpower's Untempered Crusader*. Maxwell AFB, AL: Air University Press, 1997.

Carter, Kit C., and Robert Mueller. *U.S. Army Air Forces in World War II: Combat Chronology 1941–1945*. Washington, DC: Center for Air Force History, 1991.

Chaliand, Gerard, ed. *The Art of War in World History: From Antiquity to the Nuclear Age*. Berkeley: University of California Press, 1994.

Chun, Clayton K. S. *Aerospace Power in the Twenty-First Century: A Basic Primer*. Colorado Springs, CO: United States Air Force Academy, 2001.

Churchill, Winston S. *Road to Victory*. London: Heinemann, 1966.

Clodfelter, Mark. *Beneficial Bombing: The Progressive Foundations of American Air Power, 1917–1945*. Lincoln: University of Nebraska Press, 2010.

Cody, James R. *AWPD-42 to Instant Thunder: Consistent, Evolutionary Thought or Revolutionary Change?* Maxwell AFB, AL: Air University Press, 1996.

Copp, Dewitt S. *A Few Great Captains: The Men and Events That Shaped the Development of U.S. Air Power*. Garden City, NY: Doubleday, 1980.

———. *Forged in Fire: Strategy and Decisions in the Air War over Europe, 1940–45*. Garden City, NY: Doubleday, 1982.

Craven, John, ed. *The 305th Bomb Group in Action: An Anthology*. Burleson, TX: 305th Bombardment Group (H) Memorial Association, 1990.

Craven, Wesley F., and James L. Cate, eds. *The Army Air Forces in World War II*. Vol. 1: *Plans and Early Operations, January 1939 to August 1942*. Chicago: University of Chicago Press, 1948.

———. *The Army Air Forces in World War II*. Vol. 2: *Europe: Torch to Pointblank, August 1942 to December 1943*. Chicago: University of Chicago Press, 1949.

de Marmont, Duc de Ragusa, Auguste Frédéric Louis Viesse. *The Spirit of Military Institutions; or, Essential Principles of the Art of War*. Translated by Henry Coppee. Philadelphia: Lippincott, 1862.

Douhet, Giulio. *Command of the Air*. Washington, DC: Air Force History and Museums Program, 1998.

DuPre, Flint O. *U.S. Air Force Biographical Dictionary*. New York: Franklin Watts, 1965.

Edkins, Craig R. *Anonymous Warrior: The Contributions of Harold L. George to Strategic Air Power*. Maxwell AFB, AL: Air Command and Staff College, 1997.

Emme, Eugene M. *The Impact of Air Power: National Security and World Politics*. Princeton: D. Van Nostrand, 1959.

Ennels, Jerome A. *The Wisdom of Eagles: A History of Maxwell Air Force Base*. Montgomery, AL: Black Belt, 1997.

Faber, Peter R. "The Development of US Strategic Bombing Doctrine in the Interwar Years: Moral and Legal?" *Journal of Legal Studies* (1996/1997). Accessed November 11, 2017, www.au.af.mil/au/awc/awcgate/interwar/faberdbd.htm.

Field Service Regulations. Part 1: *Operations, 1909*. London: HMSO, 1914.

Finney, Robert T. *History of the Air Corps Tactical School, 1920–1940*. Maxwell AFB, AL: USAF Historical Division, 1955.

Freeman, Roger A. *Mighty Eighth War Diary*. Minneapolis: Motorbooks, 1970.

Frisbee, John L. *Makers of the United States Air Force*. Washington, DC: Office of Air Force History, 1987.

Fuller, J. F. C. *The Reformation of War*. London: Hutchinson, 1923.

Futrell, Robert F. *Ideas, Concepts, Doctrine: Basic Thinking in the United States Air Force, 1907–1960*. Maxwell AFB, AL: Air University Press, 1989.

Gaston, James C. *Planning the American Air War: Four Men and Nine Days in 1941.* Washington, DC: National Defense University Press, 1982.

Getz, Lowell. *"Mary Ruth" Memories of Mobile . . . We Still Remember: Stories from the 91st Bombardment Group (Heavy), 8th Air Force, World War II.* 2001. Accessed October 28, 2017, http://www.91stbombgroup.com/mary_ruth/indexmaryruth.html.

Giffard, Hermione. "Engines of Desperation: Jet Engines, Production and New Weapons in the Third Reich." *Journal of Contemporary History* 48, no. 4 (2013): 821–44.

Gilbert, Martin. "Churchill Proceedings—Churchill and Bombing Policy." Fifth Churchill lecture. George Washington University, October 18, 2005. Accessed May 30, 2018, www.winstonchurchill.org/publications/finest-hour/finest-hour-137/churchill-proceedings-churchill-and-bombing-policy/.

Goltz, Baron Von Der. *The Nation in Arms.* Translated by Philip A. Ashworth. London: W. H. Allen, 1887.

Gorrell, Edgar S. *The Measure of America's World War Aeronautical Effort.* Norwich, VT: Norwich University, 1940.

Grandstaff, Mark. "Muir Fairchild and the Origins of Air University, 1945–46." *Air Power Journal* (Winter 1997).

Green, Murray. "Interview, LtGen Harold L. George." West Los Angeles, March 16, 1970. Air Force Historical Research Agency.

Greer, Thomas H. *The Development of Air Doctrine in the Air Arm, 1917–1941.* USAF Historical Study 89. Maxwell AFB, AL: Air University Press, 1955.

Griffen, H. Dwight, et al. *Air Corps Tactical School: The Untold Story.* Maxwell AFB, AL: US Government Printing Office, 1995.

Griffith, Charles. *The Quest: Haywood Hansell and American Strategic Bombing in World War II.* Maxwell AFB, AL: Air University Press, 1999.

Guderian, Heinz. *Panzer Leader.* New York: Da Capo, 1952.

Hansell, Haywood S. *The Air Plan That Defeated Hitler.* Atlanta: Higgins-McArthur, 1972.

———. "American Air Power in World War II." Maxwell AFB, AL: USAF Historical Research Center, n.d., K112.3-2.

———. "Interview of Maj Gen Haywood S. Hansell, Jr." April 19, 1967. Maxwell AFB, AL: United States Air Force Oral History Program AFHRA K239.0512-628.

———. *The Strategic Air War against Germany and Japan.* Washington, DC: Office of Air Force History, 1986.

Hastings, Max. *Bomber Command.* London: Pan Books, 1999.

Head, William P. *Every Inch a Soldier: Augustine Warner Robbins and the Building of US Airpower.* College Station: Texas A&M University Press, 1995.

Holmes, Harry. *Avro Lancaster: The Definitive Record.* London: Butler & Tanner, 1997.

Hucker, Daniel. *Public Opinion and the End of Appeasement in Britain and France.* Farnham, UK: Ashgate, 2011.

Hurley, Alfred F. *Billy Mitchell: Crusader for Air Power.* Bloomington: Indiana University Press, 1975.

Irons, Roy. *Hitler's Terror Weapons: The Price of Vengeance.* London: HarperCollins, 2002.

Johnston, Timothy. *Being Soviet: Identity, Rumour, and Everyday Life under Stalin, 1939–1953.* Oxford: Oxford University Press, 2011.

Jones, F. C. *Japan's New Order in East Asia: Its Rise and Fall, 1937–45.* New York: AMS, 1978.

Jones, H. A. *The War in the Air: Being the Story of the Part Played in the Great War by the Royal Air Force*. Vol. 3. Oxford: Clarendon, 1931.

———. *The War in the Air: Being the Story of the Part Played in the Great War by the Royal Air Force*. Vol. 6. Oxford: Clarendon, 1937.

Jones, R. V. *The Wizard War: British Scientific Intelligence, 1939–1945*. New York: Coward, McCann & Geoghegan, 1978.

Kaplan, Fred. *The Wizards of Armageddon*. Stanford, CA: Stanford University Press, 1983.

Kennett, Lee. *The First Air War, 1914–1918*. Toronto: Free Press, 1991.

Kreis, John F. *Piercing the Fog*. Washington, DC: Air Force History and Museums Program, 1996.

MacIsaac, David. *Strategic Bombing in World War II: The Story of the United States Strategic Bombing Survey*. New York: Garland, 1976.

Mastny, Vojtech, "Stalin and the Prospects of a Separate Peace in World War II." *American Historical Review* 77, no. 5 (December 1972): 1365–88.

Mathewson, Eric S. *The Impact of the Air Corps Tactical School on the Development of Strategic Doctrine*. Maxwell AFB, AL: Air Command and Staff College, 1998.

Maurer, Maurer. *The U.S. Air Service in World War I*. Vol. 2. Washington, DC: Office of Air Force History, 1978.

———. *The U.S. Air Service in World War I*. Vol. 4. Washington, DC: Office of Air Force History, 1979.

McFarland, Stephen. *America's Pursuit of Precision Bombing, 1910–1945*. Washington, DC: Smithsonian Institute Press, 1995.

Meilinger, Phillip S. *Bomber: The Formation and Early Years of Strategic Air Command*. Maxwell AFB, AL: Air University Press, 2012.

———. *The Paths of Heaven: The Evolution of Airpower Theory*. Maxwell AFB, AL: Air University Press, 1997.

Miller, Roger G. *Billy Mitchell: Stormy Petrel of the Air*. Washington, DC: Air Force History and Museums Program, 2004.

Millman, Brock. "British Home Defence Planning and Civil Dissent, 1917–1919." *War in History* 5 (1998): 204–32.

Mitchell, William. "Notes on the Multi-motored Bombardment Group Day and Night," 1923. Maxwell AFB, AL: Fairchild Special Collection 358.4 U58n, Air University Library.

———. *Winged Defense: The Development and Possibilities of Modern Air Power, Economic and Military*. New York: Dover, 1988.

Morris, Joseph. *The German Air Raids on Great Britain, 1914–1918*. London: Sampson Low, Marston, 1925.

Morrison, Wilbur. *The Incredible 305th: The "Can Do" Bombers of World War II*. New York: Berkley, 1962.

Muller, Richard. *The German Air War in Russia*. Baltimore, MD: Nautical & Aviation, 1992.

Munson, Kenneth. *Aircraft of World War II*. New York: Doubleday, 1962.

Murray, Williamson. *German Military Effectiveness*. Mt. Pleasant, SC: Nautical & Aviation, 1992.

———. "Reflections on the Combined Bomber Offensive." *Militargeschichtliche Mitteilungen* 51 (1992): 73–94.

———. *Strategy for Defeat: The Luftwaffe, 1933–45*. Honolulu: University Press of the Pacific, 2002.
Murray, Williamson, and Allan Millet, eds. *Military Innovation in the Interwar Period*. Cambridge: Cambridge University Press, 1996.
Nalty, Bernard C. *Winged Shield, Winged Sword: A History of the United States Air Force*. Washington, DC: Air Force History and Museums Program, 1997.
Neufeld, Michael. "Hitler, the V-2, and the Battle for Priority, 1939–1943." *Journal of Military History* 57, no. 3 (July 1993): 511–38.
91st Bombardment Group Official Diary, April 1943. AFHRA GP-91-HI 1. Maxwell AFB, AL: Air Force Historical Research Agency, April 30, 1943.
O'Brien, Phillips. *How the War Was Won*. Cambridge: Cambridge University Press, 2015.
Ose, Dieter. "Rommel and Rundstedt: The 1944 Panzer Controversy." *Military Affairs* 5, no. 1 (January 1986).
Overy, Richard. *The Bombers and the Bombed: Allied Air War over Europe, 1940–1945*. New York: Viking, 2013.
———. *Why the Allies Won*. New York: Norton, 1995.
Paine, S.C.M. *The Wars of Asia, 1911–49*. New York: Cambridge University Press, 2012.
Pardini, Albert. *The Legendary Norden Bombsight*. Atglen, PA: Schiffer, 1999.
Parks, W. Hays. "'Precision' and 'Area' Bombing: Who Did Which, and When?" *Journal of Strategic Studies* 18, no. 1 (1995): 145–74.
Parramore, Woody. "The Combined Bomber Offensive's Destruction of Germany's Refined-Fuels Industry." *Air & Space Power Journal*, March–April 2012, 72–89.
Purtee, Edward O. *History of the Army Air Service, 1907–1926*. Wright-Patterson AFB, OH: Historical Office, 1948.
Raleigh, Walter. *The War in the Air: Being the Story of the Part Played in the Great War by the Royal Air Force*. Vol. 1. Oxford: Clarendon, 1922.
Rommel, Erwin. *The Rommel Papers*. New York: Da Capo, 1953.
Royston, Mark W. *The Faces behind the Bases: Short Biographies of Those for Whom Military Bases Are Named*. Bloomington, IN: iUniverse, 2009.
Severs, Hugh G. *The Controversy behind the Air Corps Tactical School's Strategic Bombardment Theory: An Analysis of the Bombardment versus Pursuit Aviation Data between 1930–1939*. Maxwell AFB, AL: Air Command and Staff College, 1997.
Sherman, Don. "The Secret Weapon." *Air & Space Smithsonian*, February–March 1995, 78–87.
Sherman, William. *Air Warfare*. Maxwell AFB, AL: Air University Press, 2002.
Sherry, Michael S. *The Rise of American Air Power*. New Haven: Yale University Press, 1987.
Shiner, John F. *Foulois and the U.S. Army Air Corps 1931–1935*. Washington, DC: Office of Air Force History, 1983.
Shrader, Charles R. *History of Operations Research Analysis in the United States Army*. Vol. 1: *1942–1962*. Washington, DC: United States Army, 2006.
Stubbs, David. "A Blind Spot? The Royal Air Force (RAF) and Long-Range Fighters, 1936–1944." *Journal of Military History* 78 (April 2014): 673–702.
Testimony Presented by Major Donald Wilson, Captain Robert Olds, Captain Harold Lee George, Captain Robert M. Webster, 1st Lieutenant K. N. Walker before the Federal Aviation Commission, Washington, D.C. Maxwell AFB, AL, Air Force Historical Agency 248.121-23.

Thom, Walter W. *The Brotherhood of Courage: The History of the 305th Bombardment Group (H) in World War II*. New York: Martin Cook, 1986.

303rd Bombardment Group Official Diary, Jan–Dec 1943, April 17, 1943. AFHRA GP-303-HI. Maxwell AFB, AL: Air Force Historical Research Agency.

305th Summary of Events, 1 November 1942 to 31 December 1943, December 1943. AFHRA GP-305-HI. Maxwell AFB, AL: Air Force Historical Research Agency.

306th Bombardment Group History, Mar 1942–Jun 1945. AFHRA GP-306-HI. Maxwell AFB, AL: Air Force Historical Research Agency.

306th Bombardment Group History, Jan 1943 to Dec 1943. GP306-HI (Bomb). Maxwell AFB, AL: Air Force Historical Research Agency.

Tooze, Adam. *Wages of Destruction: The Making and Breaking of the Nazi Economy*. New York: Penguin, 2006.

Trenchard, Hugh. "Memorandum by the Chief of the Air Staff for the Chiefs of Staff Sub Committee on the Object of an Air Force, 2nd May 1928." In *The Strategic Air Offensive against Germany, 1939–1945*, vol. 4, *Annexes and Appendices*. Edited by Charles Webster and Noble Frankland. London: Her Majesty's Stationery Office, 1961.

———. "Report on the Independent Air Force." *Tenth Supplement to London Gazette*, January 1, 1919.

Twing, Stephen W. *Myths, Models & U.S. Foreign Policy: The Cultural Shaping of Three Cold Warriors*. London: Lynne Rienner, 1998.

United States Strategic Bombing Survey. *Aircraft Division Airframes Plant Report No. 5: Focke-Wulf Aircraft Inc, Bremen, Germany*, Air Frames Branch, September 26, 1945. Library of Congress, D785.U6 No. 10A.

Walker, Kenneth N. "Driving Home the Bombardment Attack." *Coast Artillery Journal* 73, no. 4 (October 1930).

Weal, John. *BF109 Defence of the Reich Aces*. New York: Bloomsbury, 2012.

Werrell, Kenneth. *Blankets of Fire: U.S. Bombers over Japan during World War II*. Washington, DC: Smithsonian Institute Press, 1996.

———. "The Strategic Bombing of Germany in World War II: Costs and Accomplishments." *Journal of American History* 73, no. 3 (December 1986): 703–13.

Williams, George. "The Shank of the Drill: Americans and Strategical Aviation in the Great War." *Journal of Strategic Studies* 19, no. 3 (September 1996): 381–431.

Williams, Gordon. "A Progressive History of the Air Corps Egg Layers." *Flying Magazine* 19, no. 5 (November 1936).

Wilson, Donald. "Origin of a Theory for Air Strategy." *Aerospace Historian* 18, no. 1 (March 1971).

———. *Wooing Peponi: My Odyssey through Many Years*. Monterey, CA: Angel, 1973.

Index

Page numbers in italics refer to illustrations.

accuracy, of bombing, 137–38; correction of bombing inaccuracy, 118–27, *122*; Norden bombsight and, 22, 24, 117, 122, 124–25; RAF in World War II, 13; sighting shot and, 123, 126, 132; in Spain and China, 133–35; USAAF in World War II, 207; wind and, 132, 133, 220

ACTS (Air Corps Tactical School), 6, 18, 19–21, 85; Air Force Course, 29, 33, 46, 99; Air Tactics and Strategy Department, 20, 28, 29, 47; Austin Hall, *27*; Bombardment Section, 122, 126–27; bombing records at, *122*, 126; contribution to victory in World War II, 223–24; faculty and students, 26–29, *27*, *28*, 98, 117, 178, 193, 199; HADPB concept developed at, 5; truncated curriculum (1939–1940), 195; World War I influence on, 177

Afghanistan, 138

Agincourt, Battle of (1415), 67

air alert, 111, 112

air bases (airdromes), 56, 187, 190; B-29 bases in China, 219–20; in England (World War I), 102; in England (World War II), 1, 201; German (World War I), 106; offensives against, 55, 97

Air Corps, US, 49, 105, 223; Air Corps Board, 130–31, 137; early bomber models, 21, *21*, *22*, *23*; expansion in anticipation of war, 194; Norden bombsight and, 22, 24, *24*. See also ACTS

air defense, 3, 20, 53; British Chain Home system, 115; German, 204–5; improvements in, 6, 7; offensive forces neutralized by, 11; over wide territory, 20; in World War I, 6–7, 8, 102, 103. See also anti-aircraft artillery [AAA] (flak); pursuit aircraft

air power, 18, 29, 35, 44, 47, 48; defensive, 55; equality with land and sea power, 14, 36; German, 50–51, *51*, 58; inability to win a war alone, 222–23; independent use of, 9, 35; Italian, 56, 58; mobilization of, 50; offensive nature of, 10, 11, 13, 19, 45, 85; principles of war and, 69, 70; strategical conception of, 48, 49; victory through disruption of economic system, 30–31, 33, 42, 43, 44–45; in World War I, 6, 8, 101

air superiority, 6, 10; attritional battle for, 45; German failure to attain in Battle of Britain, 198; HADPB and, 209–11, 212, 213; in Pacific War, 222; as prerequisite for Operation Overlord, 196–97

Air War College, 125

air warfare, 32, 50, 141; advantages peculiar to, 82; decision making in, 30; definition of, 51; relation to traditional warfare, 181–83; strategy of, 73, 74

290   Index

Air Warfare (Sherman, 1926), 19
Alabama, USS, practice bombing of (1921), *15*, 121
Albatross fighters, German, 102, 103, 106
allies, 182, 183, 187, 193; Grand Alliance in World War II, 213–14; US–British CBO, 197, 203–4, 213–14; US–Soviet cooperation against Japan, 190–92
Annual Bombing and Gunnery Matches, 124, 125
anti-aircraft artillery [AAA] (flak), 1, 53, 87; effectiveness of, 120; flying formations in defense against, 90, 93–97, *95*; German, 213, 215, 216; placement of, 163; sighting shots employed by, 132; "suicide altitudes" and, 123; in World War I, 106
APQ-13 radar bombing system, 218–19, 221
area bombing, 20, 198, 212, 215, 221
Army Air Service, US, 14, 15, 19
Army Command and General Staff College, 26, 122–23
Army Industrial College, 26
Army-Navy Exercises (Langley Field, 1938), 122
Army Reorganization Act (1920), 14
Army Training Regulations, 19
Army War College, 26, 100, 123
Arnold, General Hap, 198–99, *199*, 219, 220–21
attrition (losses), of bombers, 97; from accidents and mechanical trouble, 106; Bremen raid (1943), 2; Schweinfurt and Regensburg raids (1943), 98, 204; Stuttgart raid (1943), 204; tactical offense–defense ratio and, 98–99, 103–13, *104*, *107*
Austria, 57
Austro-Prussian War (1866), 73
AWPD-1 (Air War Plans Division), 31, 195, 209, 220; ACTS former faculty in, 199; methodology of, 202; strategic objective of, *200–201*
AWPD-42, 202, 220

B-2 Condor (Curtiss) bomber, 21, *22*
B-10 (Martin) bomber, 21, *23*, 24, 119
B-17 [Flying Fortress] (Boeing) bomber, 6, 21, 29, 98, 138, 179, 196; 92nd Bombardment Group, *2*; attrition in air campaign against Germany, 204, *206*; B-17G, *26*; bomb capacity of, 52, 119; Bremen raid (1943), 1–2, *3*; H2X radar bombing system, 212, 215, 219; *Memphis Belle*, *3*; Norden bombsight in, 24, 29, 138, 139, 179; production of, 209; range of, 217; Schweinfurt raid (1943), *205*; YB-17 "test" aircraft, 24, *25*
B-18 Bolo (Douglas) bomber, 24, *25*, 119, 134, 136
B-24 bomber, 196, 209, 217
B-29 (Boeing) bomber, 5, 196, 217–23, *218*
Baldwin, Stanley, 13
ball bearing plants, as targets, 204, 219
ballistic wind effect, 132, 133
Berlin, bombing raids on, 204
blind flying and landing, 34, 43
bombardment, 19, 69, 87–89; altitudes for, 117, *122*, 123, 132; best defense against, 97; bombing records, 126, 127, 128–29, 130, 131, 136; capacity and efficiency of, 54; courses on, 29; "down the groove" method of approach, 125; equipment and technique, 129; German bombardment planes (1938), 50–51, *51*; as supplement to artillery barrage, 133. *See also* area bombing; HADPB
bomb damage, 13, 34; Focke-Wulf plant (Bremen, Germany), 2, 3; from German airships in World War I, 52; possible results of air attack on New York City, 174–77, *175*; in World War I, 8, 54
"Bomber Mafia," 28, 29, 31, 195, 199, 220
bombers: bomb capacity of, 89; flying formations, 87, 90–97, *92*, *95*; long-range fighter escorts, 2; radius of action, 48, 89, 111, 187, 201

Bosnia, 138
Britain, Battle of (1940), 198
British Field Service Regulations, 60, 61
Bulge, Battle of the (1944), 197, 211
Bureau of Ordnance, 22
Butt Report (1941), 13, 198, 213, 215

Cairo Conference (1943), 219
Cannae, Battle of (216 B.C.), 64, 67
Caproni, Giovanni, 9, 11, 15
Casablanca Conference (1943), 208
CBO (Combined Bomber Offensive), 197, 203–4, 213–14
Chamberlain, Neville, 58
Chennault, Claire, 27, 28, 29
Chiang Kai-shek, 219
China, 113–14, 133–34, 151; B-29 bases in, 219–20; civilian morale under Japanese air attack, 142, 177; "Open Door" policy of United States and, 76–77
chokepoints (bottlenecks), 45, 151
Churchill, Winston, 199, 213, 214
civilian population, 20, 30, 79–80, 189, 192; civilian morale as decisive factor, 78, 81, 165; as collateral damage, 12; direct attack on, 140–43; in Douhet's theory, 9–10, 16; "home front" as weak link, 62; loss of enemy population's will to resist, 32; in Mitchell's theory, 17; morale of civilians, 8, 11, 12–13; in Trenchard's theory, 11–13, 16; vulnerability to air power, 81. *See also* will to resist, enemy's
Clausewitz, Carl von, 75, 78, 80
Clay, Lieutenant Commander J. P., 135, 136
COA (Committee for Operations Analysts), 219
Cold War, 10, 178, 224
combat zone, 94–95, 96
combined arms warfare, 5, 9, 197, 209, 223
Command of the Air (Douhet, 1921), 9, 11
communications technologies, 10, 89
continental powers, 8–9, 180

Coolidge, Calvin, 18
cooperation, principle of, 59, 65–66
counteroffensive, 99, 185; German counteroffensive in Battle of the Bulge, 197; Soviet counteroffensive in World War II, 216; in World War I, 106
Crécy, Battle of (1346), 67
Czechoslovakia, 56, 57

D-1 bombsight, 123, 124
D-4 bombsight, 123, 124, 125, 134
Daladier, Edouard, 57, 58
defender-to-bomber ratio (1.5 to 1), 30
desert terrain, 138
DH-9 airplanes, British, 102, 105, 106, 109
dispersed column (dispersed group formation), 93, 94, 95, 96
dispersed production facilities, 3, 193, 196
dogfights, in World War I, 6
Dornier 177 (German bomber), 114
Douhet, General Giulio (Italian Army), 6, 9–11, 16, 19; *Command of the Air*, 9, 11; on direct bombing of cities, 177–78
Du Vernois, Verdy, 73

economy of force, 59, 61, 63
Eighth Air Force, 1, 196–97; accompanied by fighter escorts, 209–10; attrition of bombers, 2; in CBO (Combined Bomber Offensive), 203; Fighter Command, 2; P-47 (Republic) fighters, *210*; Pointblank Directive and, 204, 215
Eisenhower, General Dwight D., 208
electric power, 157–62, 168, 173–76, *175*, 178, 189
Employment of Aviation in Naval Warfare, The, 55
Ethiopia, Italian conquest of, 55, 83, 193; bombing of civilians, 142; Italian ground and air forces in, 186–87

Fairchild, Major Muir, 28–30, 45, *47*, 58, 85, 139–40; on choice of bombing

292   Index

Fairchild, Major Muir (*cont.*)
  economic or military targets, 180, 192–93; on critical requirements of industrial state, 178
"feet dry," 1
Foch, Marshal Ferdinand (French Army), 61
Focke-Wulf: Fw-190s, 2, 3; plant in Germany, 1, 2
Fokker fighters, German, 102, 103
France, 55, 56, 182; colonial empire of, 76, 147, 148; conservatism in military thinking, 67–68; economic vulnerability of, 148, 151; food production, 150; French army regulations, 61; geostrategic concerns and, 8–9; German defenses in, 212; Maginot Line defenses, 57; Munich agreement and, 56–57; preparations for protection of civilian population, 141
Frederick the Great, 82
Fuller, Major General J. F. C. (British Army), 60, 61
"Fundamental Principles of Employment of the Air Service" (Regulation 440-15, 1926), 19, 52

George, Colonel Harold, 27, 31, 34, 44–45, 85; Air Force Course, 29, 33; "Bomber Mafia" and, 26, 28; as chief of USAAF Air War Plans Division, 195, 198–99, *199*; contributions to US victory plan in World War II, 194; on defeat of enemy's will, 58–59
geostrategy, 8–9
German Air Force. *See* Luftwaffe
Germany, 29, 31, 72, 138, 182; air warning system, 10, 204, 207; collapse of economy in World War II, 179; concentrated economic structure of, 139; conservatism in military thinking, 68; economic collapse without surrender in World War II, 196; economic vulnerability of, 178; fight for national prosperity, 76; food production and consumption, 149; geostrategic concerns and, 8–9; lack of colonial empire, 76, 83; lack of raw materials, 152; Nazi control of population, 193, 196, 208, 209; in state of siege during World War I, 41; synthetic fuel production, 211; US–British CBO (Combined Bomber Offensive) against, 203–4, 213–14
GHQ Air Force, 51, 52, 105, 119, 190; effective strength of, 131, 136; Training Directives from, 126
Goering, Hermann, 15
Goltz, Lieutenant General Colmar Freiherr von der, 100–101
Gorrell, Major Edgar, 7, 8, 20
Gotha Bombers, 7
grain supplies, as target, 17
Great Britain: aircraft produced in World War I, 50; air warning system, 10, 198; colonial empire of, 75–76, 147, 148; concentrated economic structure of, 139; economic vulnerability of, 148, 151, 178; effect of German submarines on, 41, 49, 55, 64, 68; European balance of power and, 57; food production and consumption, 149; geostrategic concerns and, 9; German air raids in World War I, 6–7, 52–54, 155; importance of London to, 165; preparations for protection of civilian population, 140–41; US bombers based in, 201
Great Depression, 19, 29
ground alert, 111, 112–13
ground forces (armies, land power), 35, 47–48, 63, 72, 76, 85; defeat of enemy armed force as intermediate objective, 79; infantry supported by other ground arms, 88; limitations of, 81–82; object of war and, 39–40; occupation of enemy's territory by, 76, 79, 181; primary strategic objective of, 84, 182; rapidly moving enemy force, 186; war plans for, 146

Index   293

groups, of bombers, 88, *95*, 137
Guam, 17, 220
Gulf War (1991), 224

HADPB (high-altitude daylight precision bombing), 5, 20, 29–32, 179, 194; achievements of, 208–12; ACTS curriculum dominated by, 28; alternative explanations for utility of, 213–17; bomber formations, 90–97, *92*, *95*; centrality to USAAF war effort, 195, 199; diversion of German war resources from Eastern Front and, 215–16; genesis of, 6; German synthetic fuel production destroyed by, 200, 211; impact on German war production, 214–15; against Japan, 217–24; as prewar rationale for long-range bomber force, 196, 209; tested in World War II, 204–8, *205*, *206*
Haig, Field Marshal Douglas (British Army), 11
Hamburg, bombing of (1943), 204
Hansell, Captain Haywood, 28, 29–30, 72, *73*, 85; in AWPD (Air War Plans Division), 199, 202, 220; contributions to US victory plan in World War II, 194; fired by Arnold, 220–21; resistance to attacks on Japanese economy, 222–23; in Strategic Air Intelligence Office, 202
Harmon, Lieutenant Colonel H. R., 125
Harris, Air Chief Marshal Sir Arthur (RAF), 203
Hawaii, 17, 122, 187
Hiroshima, atomic bombing of, 5, 222
Hitler, Adolf, 57, 82, 119, 198, 212; Grand Alliance against, 213, 214; Munich appeasement of, 13; revenge weapons and, 217
Hopkins, Major Frederick, 30, 98–99, 115

Independent Force (IF), British, 7–8, 99, 101, 109; organization of, 102; pursuit-to-bomber ratio formulated by, 104. *See also* RAF (Royal Air Force)

initial point (IP), for bombing runs, 2
Instant Thunder campaign (1991), 224
Invasion 2nd, loss of, 2
Iraq, 138
Italy: conquest of Ethiopia, 55, 83, 142, 186–87, 193; defiance of British navy in 1930s, 55–56; economic vulnerability of, 178; fascism in, 11; fight for national prosperity, 76; food production, 150; geostrategic concerns and, 9; lack of raw materials, 152; in World War I, 9

Japan, 5, 10, 31, 138; atomic bombings of, 5, 222, 223; bombing operations in China, 113–14, 133–35, 142; collapse of economy in World War II, 179; concentrated economic structure of, 139; conflict with American "Open Door" policy in China, 76–77; conquest of Manchuria, 83; economic vulnerability of, 148, 178; HADPB campaign against, 196, 217–24, *218*, *221*; island-hopping advance across Pacific toward, 187; lack of raw materials, 152; Mannatsui Bombing School, 113; prospect of war with United States (1920s), 17; US–Soviet alliance against, 190–92
Javelin formation, *92*

Katiuska (Soviet bomber), 114
Keystone airplane, 119, 124, 125, 134
Kitchener, Field Marshal Horatio Herbert, 101
Korea/Korean War, 115, 138, 193, 224
Kosovo, 138
Kursk, Battle of (1943), 214, 216
Kuter, Captain Laurence, 28, 30, 31, 117, *118*; in AWPD (Air War Plans Division), 199, 202; contributions to US victory plan in World War II, 194
Kuwait, 138

Lancaster bomber, British, 198, 215
Langley Field, Virginia, 19, 20, 122, 123

294 Index

LeMay, Major General Curtis, 220, 221, *221*, 222, 223
Libya, 138
Lloyd George, David, 7
London, 165
longbow, English, 67, 72
Luftwaffe (German air force), 1, 58, 203; in Battle of Britain, 198; British and French fear of, 13; carpet bombing of Rotterdam, 197–98; losses inflicted on Eighth Air Force, 204; losses suffered by, 3, 5, 209–10; operations restricted by loss of fuel production, 211; Soviet counteroffensive and, 216; strength (1938), 50–51, *51*
Lynch, Major General George A., 68–69

machine guns: battle of wills and, 80; defensive warfare strengthened by, 41; mounted on bombers, 24, 89, 90, 91, 105; power in World War I, 68–69, 72; of pursuit aircraft, 91, 92
Manhattan Project, 218
Marathon, Battle of (490 B.C.), 66
"marginal force," 185
Marmont, Marshal, 61
mass, principle of, 59, 63–64
Maxwell Field, Alabama, 5, 20, 26, 28
MB-2 (Martin) bomber, *15*, 21, *21*
McNair Board, 121–22
Memphis Belle B-17F, *3*
Messerschmitt Me-109s and Me-110s, 2
Milling, Major Thomas, 19
Mitchell, Brigadier General William "Billy," 6, 11, 13–19, *14*; court-martial of, 18, *18*; *Winged Defense*, 18
Morrow Board, 18, 19
Mosquito light bombers, British, 215
movement, principle of, 59, 63
Munich agreement (1938), 13, 56, 58, 69
Mussolini, Benito, 10, 82, 119

Nagasaki, atomic bombing of, 5, 222
Napoleon I, 61, 82
national economic structure, 30, 180, 183–84, 224; dependency on chains of operations, 145–46; enemy will to resist and, 143; means of waging war and, 143–45; petroleum production, 152–54; as primary strategic objective, 185, 192; US "vital industrial area," 155–64. *See also* raw materials; vital and vulnerable nodes
national security, 36, 37
Nation in Arms, The (Goltz, 1883), 100
Naval War College, 26, 55
navigation, 13, 34, 43
Navy Department, 18
neutral countries, 142, 143
New York City, industrial area of, 164–67, 176–77; electric power systems, 173–76, *175*, 178; food supply, 167, 171–72, 176; transportation networks, 167–68, 172–73, 177, 178; water supply, 168–71, 176, 178
night flying/bombing, 44, 198
Norden, Carl L., 22
Norden Mark I bombsight, 6, 22, 117, 122, 138; accuracy of, 22, 24, 117, 122, 124–25, 207; cloud cover preventing use of, 215; comparison with D-1 and D-4 bombsights, 124–25, 134; development of, 24, *24*, 26; HADPB and, 29; mounted in B-17s, 24, 29, 138, 139, 179; reduction of probability error and, 127
Normandy ground invasion, 5
Norstad, Brigadier General Lauris, 220, *221*
nuclear weapons, 10

objective, principle of the, 62–63, 185, 189, 192
occupation, of enemy's territory, 39–40, 72, 79
offense–defense balance, 5, 41, 207
offensive, principle of the, 63, 68, 72
Ostfriesland, sunk in aerial demonstration (1921), 15, *16*, 19, 121
Ottoman Empire, 9

P-26 (Boeing) pursuit, *23*

Index 295

P-47 (Republic) fighters, 2, *210*
P-51 Mustang fighters, *211*
Pattinson, Major L. A. (British Army), 103
Pee Dee River bridge bombing (1927), 122
petroleum production, 5, 152–54, 178, 197, 211
Philippines, 17
PN-9 sea plane, 18
Pointblank Directive, 204, 215
Poland, 64, 65, 69, 72, 197
Principles of War, The (Foch), 61
pursuit aircraft, 10, *23*, 28; diving, 100; flying formations in defense against, 90, 91–97, *92*; friendly, 94–95, 96; German, *51*; methods of pursuit operations, 111; ratio to bombers and attrition rate, 30, 99, 103–12, *104*; time-fused fragmentation bombs employed by, 93

"racial unity," national interest and, 29, 72, 75, 77
radar, early-warning, 1, 5, 198; in Chain Home air defense system, 115; Douhet's theory and, 10; German, 204, 207; impact on offense-defense balance, 117
radar bombing, 212, 215
RAF (Royal Air Force), 15, 198; air bases, 1; in Battle of Britain, 115; "Circus" daylight raids (France, 1940–1941), 210; creation of (1918), 7; in interwar period, 11–13; Oboe radio transponder system, 215; Spitfires, 2; in World War I, 7, 8
railroads, 133, 154–55; of New York industrial area, 167, 168, 169, 172–73; Trans-Siberian Railway, 190, 192
raw materials, 40, 75, 76, 145, 178; Japan's lack of, 148; strategic materials, 150–51; US abundance of, 147, 148–50
rear gunners, 89, 90
reconnaissance planes, 1, 6, 185
"Recording, The" (Harmon), 125
Regia Aeronautica (Italian air force), 58
Rommel, General Erwin, 212

Roosevelt, Franklin D., 56, 198–99, 208, 209, 213, 219, 220
Route Column formation, 95
Royal Flying Corps, 7, 53. *See also* RAF (Royal Air Force)
Ruhr Valley, bombing of (1943), 204, 214
Rundstedt, General Karl von, 212
Russia (Soviet Union), 10, 57, 149, 196; continental size of, 139, 147, 165; food production, 150; geostrategic concerns and, 8–9; lack of direct access to the sea, 147–48; US-British strategic bombing of Germany and, 213–14; war against Japan and, 190–92, 222; World War II alliance with United States, 190–92, 213–14, 216

Savoia 79 (Italian bomber), 114
sea (naval) forces, 14, 35, 50, 85; British navy, 55, 57, 58; hostile, 187; limitations of, 81–82; object of war and, 40; primary strategic objective of, 80, 84, 182; territory acquisition and, 83
searchlights, 53, 112
security (protection), principle of, 59, 64–65, 185, 190, 192
self-preservation, of nations, 37–38
Shaw, Lieutenant J. D., 134, 135
Shenandoah (dirigible), 18
Sherman, Major William, 19–20
sighting shot, 123, 126, 132
Simonds, Frank, 75
simplicity, principle of, 59, 65
slant ranges, 219, 220, 222
Smuts, General Jan Christian (British Army), 7
Smuts Committee, 7
Soviet Union. *See* Russia
Spain, air war in (1930s), 114, 133–34, 142, 143, 177
Spanish-American War, 68
speed, 22, 93; of B-17 bomber, 24, 29, 98; bombing accuracy and, 127; of flying formation, 94; of heavy and light bombers, 89; of MB-2 biplane, 21; of pursuit aircraft, 91, 96; ratio of pursuit

speed (*cont.*)
  to bomber speeds, 98, 100, 109–14, *110*; reduced by high winds, 109
Sperry Gyroscope Company, 22
squadrons, 88–89, *92*, 93, 137
Stalin, Joseph, 82, 119, 213–14
Stalingrad, Battle of (1942–1943), 214
strategic bombing theory, 3–6, 16–17, 44, 85, 98, 178, 219; British versus German strategic bombing, 198; hostile forces as primary strategic objective, 180–94; in interwar period, 8–19, 209; post–Cold War expansion of, 224; put to test in World War II, 31; Strategic Bombing Survey after World War II, 178; will of enemy population as target, 8; in World War I, 6–8, 20. *See also* HADPB
Stupntakit B17F, 1
surface forces. *See* ground forces; sea (naval) forces
surprise, principle of, 59, 64

target identification/selection, 9–10, 11, 13, 30, 138, 208; in air campaign against Japan, 219, 222; AWPD-1 (Air War Plans Division) and, 202; economic versus military targets, 180; in Mitchell's theory, 16–17; in peacetime, 146, 180; in Trenchard's theory, 12
Tentative Field Service Regulations (FM 100-5 [1939]), 62
Tiverton, Major Lord, 7
Tokyo, 165
Tokyo, firebombing of (1945), 221–22
Tooze, Adam, 214
Training Regulations, 60, 61, 126, 137; TR 440-40, 125, 128
transportation networks, 5, 152, *156*, 178; civilian workers' dependency on, 42–43; congestion of, 145; in Germany and German-controlled areas, 196, 208, 212; industrial production and, 139, 157; of Japan, 223; of military personnel and material, 12; of New York industrial area, 167–68; petroleum production and, 153; surface forces and, 83
Trenchard, Major General Hugh (British Army), 6, 7, 11–13, 14, 16, 19; as commander of Independent Force, 103; dispersed area bombing supported by, 20, 198; on worker morale as target, 7–8, 16, 20
20th Air Force, 219, 220

Udet, Ernst, 15
United States: continental size of, 139, 147, 165; dispersed economic structure of, 139; economic prosperity of, 76; geostrategic concerns and, 9; industrial capacity in wartime, 145; isolationism of interwar period, 14, 139, 178; New York industrial area, 164–77; nuclear standoff with Soviet Union, 10; petroleum production and refining in, 152–54, 178; prospect of war with Japan (1920s), 17; steel industry, 154–55; "vital industrial area" of, 155–64
USAAF (US Army Air Forces), 29, 31, 202, 204; Air War Plans Division, 195, 198–203; in CBO (Combined Bomber Offensive), 203; development of B-29 bomber, 217, 218; HADPB against Japan and, 197; lost opportunity to test HADPB, 196

V-1 and V-2 rockets (Nazi revenge weapons), 213, 217
Versailles, Treaty of, 75
Vietnam/Vietnam War, 115, 138, 193, 224
vital and vulnerable nodes, 20, 30, 45, 140, 179, 195, 196; economic paralysis from bombing of, 194; HADPB campaign against Japan and, 219; identification of, 178, 208; weather conditions in attacks on, 207

Wages of Destruction (Tooze), 214
Walker, Lieutenant Kenneth, 30, 31, 87, *88*, 98; in AWPD (Air War Plans Divi-

sion), 199; contributions to US victory plan in World War II, 194
war: as clash of wills, 77, 80; definition of, 35, 37, 75; economic interdependence of nations and, 42–43; historical methods of waging, 38–39, 42, 59–60; mechanization of, 101, 152, 188–89; new methods of waging, 36, 43–44, 81, 85; object of, 35, 38, 62–63; offense-defense balance, 5, 41, 207; principles of, 60–71, 185; reasons for occurrence of, 37; ultimate aim of, 74, 79–82, 84, 146, 164, 182, 193. *See also* air warfare
War Department, 14, 18, 52, 144, 156, 190
water supplies, as target, 17, 168–71
weather, 3, 54, 106, 109, 132; limits on daylight operations and, 1; in World War II air operations, 197, *206*, 207, 222
Weeks, John, 18
will to resist, enemy's, 32, 33, 45, 59, 183; air forces' direct attack on, 84; attack on economic structure and, 143; denial of life necessities and, 41; direct bombing of cities and, 177–78; means and will interrelated, 72, 78; object of war to break, 38, 39, 72; will to resist versus will to fight, 78–79
Wilson, Lieutenant Colonel Donald, 27–29, 58–59, *60*, 71, 85, 139
Winged Defense (Mitchell, 1925), 18
wings (organization unit of bombers), 88
World War I, 6–8, 13, 30, 99, 180; American Expeditionary Force, 14; attritional trench warfare, 101, 205; blunders produced by fixed ideas, 68–69; British Official History of the War, 52–54; Central Powers in state of siege, 79, 144; costs in lives and wealth, 41; as economic struggle, 143–44; French principle of the offensive in, 63, 68, 72; German break-through (March 1918), 186; German forces undefeated in, 40–41, 62, 72, 78; German submarine campaign, 41, 49, 55, 64; machine gun power in, 68–69, 72; numbers of aircraft produced in, 50; pursuit-to-bomber ratio, 103–12, *104*, *107*, *110*; quantity and cost of munitions fired in, 144; Zeppelin airship attacks on Britain, 6–7, 52–54, 155

World War II, 5, 22, 178; ACTS graduates in, 28; air war in Europe (1939–1941), 197–98; Allied demand for unconditional surrender, 208; attrition rate of bombers in, 115; defensive technologies in, 10; economic paralysis without immediate surrender in, 45, 179, 196; German forces interdicted by USAAF, 212; Operation Overlord, 196–97; success of bomber formations in, 98; US–British CBO (Combined Bomber Offensive), 203–4; US–Soviet alliance against Japan, 190–92

XF-38 pursuit plane, 111
XX Bomber Command, 219, 220
XXI Bomber Command, 220, 222

Y1B-9 (Boeing) bomber, 21, *23*

Zeppelin dirigibles (airships), 6–7, 52–54, 155

A Mitchell Institute Book
Aviation and Air Power
Series Editor: Brian D. Laslie

In his work *Winged Defense,* Brigadier General William "Billy" Mitchell stated, "Air power may be defined as the ability to do something in the air." Since Mitchell made this statement, the definition of air power has been contested and argued about by those on the ground, those in the air, academics, industrialists, and politicians.

Each volume of the Aviation and Air Power series seeks to expand our understanding of Mitchell's broad definition by bringing together leading historians, fliers, and scholars in the fields of military history, aviation, air power history, and other disciplines in the hope of providing a fuller picture of just what air power accomplishes.

This series offers an expansive look at tactical aerial combat, operational air warfare, and strategic air theory. It explores campaigns from the First World War through modern air operations, along with the heritage, technology, culture, and human element particular to the air arm. In addition, this series considers the perspectives of leaders in the US Army, Navy, Marine Corps, and Air Force, as well as their counterparts in other nations and their approaches to the history and study of doing something in the air.